SOCIA

ABDOLMOHAMMAD KAZEMIPUR

SOCIAL CAPITAL AND DIVERSITY

SOME LESSONS FROM CANADA

PETER LANG

Bern·Berlin·Bruxelles·Frankfurt am Main·New York·Oxford·Wien

Bibliographic information published by Die Deutsche Bibliothek
Die Deutsche Bibliothek lists this publication in the Deutsche National-
bibliografie; detailed bibliographic data is available on the Internet at
‹http://dnb.ddb.de›.

British Library Cataloguing-in-Publication Data: A catalogue record for this book
is available from The British Library, Great Britain

Library of Congress Cataloging-in-Publication Data

Kazemipur, Abdolmohammad.
 Social capital and diversity : some lessons from Canada /
Abdolmohammad Kazemipur.
 p. cm.
 Includes bibliographical references.
 ISBN 978-3-03911-710-9 (alk. paper)
 1. Social capital (Sociology)–Canada. 2. Immigrants–Canada–Social conditions–
21st century. 3. Ethnicity–Canada. 4. Canada–Ethnic relations. 5. Canada–Social
conditions–21st century. I. Title.
 HN109.3.K39 2008
 305.800971–dc22
                                                                    2008045126

Layout and Typesetting: Ingrid Pergande-Kaufmann and Till Adloff, Berlin

Cover design: Eva Rolli, Peter Lang AG

ISBN 978-3-03911-710-9

© Peter Lang AG, International Academic Publishers, Bern 2009
Hochfeldstrasse 32, Postfach 746, CH-3000 Bern 9, Switzerland
info@peterlang.com, www.peterlang.com, www.peterlang.net

All rights reserved.
All parts of this publication are protected by copyright.
Any utilisation outside the strict limits of the copyright law, without
the permission of the publisher, is forbidden and liable to prosecution.
This applies in particular to reproductions, translations, microfilming,
and storage and processing in electronic retrieval systems.

Printed in Germany

Table of Contents

Preface	7
1 Introduction	9

PART I
THE BIG PICTURE 25

2 Canada on the Global Map	27
3 The Peaks and Valleys: Canada's Social Capital over Time	37

PART II
MULTI-DIMENSIONALITY OF SOCIAL CAPITAL AND MULTI-DIMENSIONALITY OF DIVERSITY 63

4 A Close-up on the Concept: Dimensions of Social Capital	65
5 Social Capital and Regional Diversity: The Provinces	79
6 Social Capital and Immigration Status: Immigrants and the Native-born	93

PART III
DIVERSITY AND TRUST 107

7 Trust and Diversity in Cities	109
8 Trust and Immigrant Status	133
9 Trust and Ethnicity: Culture	149

10 Trust and Ethnicity: The Economic Factor   159
11 Trust and Ethnicity: Social Interaction   165
12 Trust and Religion   187

Part IV
Putting Things Together   205

13 The Voices Behind the Statistics   207
14 Wrap-up: Conclusions and Implications   219

References   227
Appendices   235

Preface

The present book is the product of five years of extensive research on social capital in Canada. The project started out as a secondary analysis of some publicly-available Canadian data from surveys conducted by Statistics Canada. Soon, it became clear that a thorough understanding of the phenomenon under study required much more information than that. As a result, the project branched out in several different directions and incorporated a much wider range of data sources: a) a switch was made from the public-use versions of Statistics Canada's data to the 'masterfile' versions (or complete raw data), available through Statistics Canada's Research Data Centres (RDCs), which provide a much deeper and elaborate picture; b) the World Values Surveys were incorporated into the study, in order to compare Canadian trends with those of some other industrial nations; 3) a separate survey was conducted to fill the gap in the existing Statistics Canada's survey data; and, 4) a qualitative component was added – including several in-depth and face-to-face interviews – in order to get the behind-the-scene dynamics of how social capital works. The result is an extremely rich picture, only some parts of which are analyzed in the current book, which has led to the surfacing of some very interesting patterns that pose some serious questions for future research.

In a very general way, and as far as social capital trends are concerned, the findings of this project point to the possibility of the presence of a so-called 'Canadian exceptionalism.' For one thing, they convey a not-so-gloomy picture for Canada; that is, Canada's general stock of social capital seems to be in relatively good shape, compared to that of many other industrial nations in Europe as well as the United States. For another, the recent concern in Europe, the United States and Australia about the negative impact of ethnic and cultural diversity on social capital seems to be less relevant for Canada. Both of these general findings are partially confirmed in this book, although a solid argument in favour of them requires more sophisticated and specifically-designed data than is currently available. I certainly hope, however, that my discussions in this book can, at least, provide some food for thought and possibly trigger a move in that direction.

If readers find in this book something worthwhile, they should give the credit to a wide range of people and institutions, without whose support and assistance the project would not have been completed. First, I benefited greatly from the high-quality work of my research assistants, Christopher Birrell and Natasha Fairweather, then undergraduate students at the University of Lethbridge. My initial nervousness about conducting a project of this magnitude in a primarily undergraduate institution like the University of Lethbridge immediately dissipated after Chris and Natasha started working on the project. Their later success – Chris was offered a job at Statistics Canada, and Natasha started her graduate studies – testify to the quality of their work.

Second, this project was made possible by the funding and other institutional support provided through many different sources. In particular, I would like to acknowledge the financial support of the *Social Sciences and Humanities Research Council of Canada*, the *Prairie Metropolis Centre,* and the RDC initiative of Statistics Canada. Needless to say that the contents of this book reflect my own positions and opinions, and not those of any of the above institutions. On top of those, I would like to express my appreciation to Dr. Christopher Nicol, Dean of Arts and Sciences, and Dr. Dennis Fitzpatrick, Vice-President of Research, at the University of Lethbridge, for their generous support of the project by making available funding, space and, more importantly, for their encouragement for the completion of the project.

Last, but not least, I would like to express my special thanks to my wife and our two sons for putting up with me and my crazy schedule while working on this project. Many pages of this book are simply substitutes for missed soccer games, camping trips, intelligent conversations, and fun times together. Their success, however, assures me that they found other, and perhaps more effective, ways to compensate for my absence.

A. K.
Lethbridge
July 2008

1 Introduction

Social capital can be defined, in the most general way, as referring to the resources embedded in social ties among individuals. One can think of numerous examples that would fit this broad definition: finding a job through a friend, receiving support from family in times of hardship, volunteering in a neighbourhood project, and donating money and blood. Obviously, this definition needs fine-tuning, but let us postpone that for later in the book and accept this as a working definition for the purpose of reviewing the history of the concept.

Social capital, as a concept rather than a term, has a relatively long history, but its most recent emergence goes back to the last two decades of the 20th century. This recent history consists of three relatively distinct periods: the early 1980s, when the concept of social capital was invented by Pierre Bourdieu (2001 [1983]); the late 1980s, when it was introduced to the English-speaking world through the works of James Coleman (1988; 1990); and, the 1990s, when it was popularized by Robert Putnam (1993; 1995; 2000). Of these three periods, the first and the last are of much more importance, and will therefore be discussed more thoroughly in the following pages.

Bourdieu's treatment of social capital was part of his much larger project of developing a comprehensive theory of social action that would cut across, and encompass, the existing social science disciplines. In this respect, his project was the social science equivalent of the attempt made by many prominent physicists to come up with a so-called 'final theory' of physics that would incorporate the existing theories in the four separate fields of gravity, electromagnetics, large and small nuclear fields (see, for instance, Weinberg, 1993). Such a comprehensive theory, Bourdieu believed, would allow the existing findings in one field to freely migrate to another:

> Sociology is the art of thinking phenomenally different things as similar in their structure and functioning and of transferring that which has been established about a constructed object, say the religious field, to a whole series of new objects, the artistic field and so on. (quoted in Bourdieu and Wacquant, 1992: 5)

As a component of such a broad framework, Bourdieu (2001[1983]) proposed what he called "a general science of the economy of practices", in which his main goal was to remove the boundaries between classical economics and other social sciences. Towards this goal, he argued that, besides what we know as 'economic capital', there exist several other types of capital, such as cultural capital and social capital, which can be converted into economic capital or be generated by it. In this context, what he meant by capital was "power" or "resources" (Granovetter and Swedberg, 2001: 111). He argued that there exists a social equivalent to the physical principle of the conservation of energy. This principle implies that, "profits in one area [of social life, such as in the economic area] are necessarily paid for by costs in another (so that a concept like wastage has no meaning in a general science of the economy of practices)," and that, "[t]he universal equivalent, the measure of all equivalences, is nothing other than labor-time (in widest sense)... " (Bourdieu, 2001 [1983]: 106).

The forms of capital that Bourdieu proposed included: cultural capital, social capital, symbolic capital, scientific capital, juridical capital, academic capital, intellectual capital, philosophical capital, linguistic capital, and political capital (Bourdieu and Wacquant, 1992). Of these, the first two have been the most frequently-used. By cultural capital, he meant the labor-time invested in a person, mostly by family members, through the process of socialization. This investment can manifest itself through particular knowledge, certain personality traits, abilities, or talents, and may later be converted into economic capital, either directly or by facilitating success in acquiring educational credentials (a form of human capital), which can then easily be converted into occupation and then income. Social capital, for Bourdieu, was associated with membership in a group "which provides each of its members with the backing of the collectively-owned capital" (Bourdieu, 2001 [1983]: 103). He argues that the benefits that accrue to the group members are the converted forms of the time, energy, and even economic capital that they had previously spent in maintaining the group's vitality and the mutual commitment of the members towards each other:

> The reproduction of social capital presupposes an unceasing effort of sociability, a continuous series of exchanges in which recognition is endlessly affirmed and reaf-

firmed. This work, ... implies expenditure of time and energy and so, directly or indirectly, of economic capital ... (p: 104)

The second period in the history of social capital was the introduction of the concept to the English-speaking world, through the works of James Coleman. In a paper entitled "Social Capital in the Creation of Human Capital" (1988), and later in his *Foundations of Social Theory* (1990), Coleman indeed provided some empirical evidence for Bourdieu's argument that different types of capital can be converted into one another by examining the influence of social capital available to children on their degree of educational achievement. It was already known that the degree of children's success at school is heavily influenced by their parents' physical capital and human capital – a concept introduced by Becker (1964) in reference to one's skills and knowledge typically acquired through formal education; what Coleman added was that this influence does not materialize without the presence of social capital – e.g., strong relationships between parents and children, reflected in the amount of time they spend with each other. Moreover, he showed that the presence of a strong communal environment would have a similar positive influence on the educational attainment of children, even in the absence of familial social capital.

Despite the numerous references to Coleman's works, they did not trigger a noticeable wave of research on social capital. This was partly due to the fact that, for him, social capital was not a new variable *per se*. Rather, for him, social capital was more of an organizing principle that acted, for the most part, as a new packaging for some concepts that were already discussed in sociological literature including his own works. Writing on the contributions of Coleman to sociology, Marsden (2005) found the large number of references to Coleman's work on social capital rather ironic as, for him, social capital served "to group together other processes he discusses..., rather than to introduce fundamentally different ones" (p: 15). And, these 'other processes' are discussed mostly under the concept of social organization.

The publication of Robert Putnam's works on social capital (Putnam, 1993; 1995; 2000) made the concept vastly popular. Through Putnam, social capital entered into the realm of rigorous empirical research and became an indispensable tool in the hands of many researchers from diverse disciplinary backgrounds, ranging from economics to political science, and from anthropology to health sciences. This sudden popularity was due mostly to the fact that Putnam's key argument was broad, straightforward, intuitively sensible, and applicable to many disciplines. In particular, in his *Bowling Alone*, Putnam (2000) incorporated so many different variables in his discussion of American social capital trends that many researchers from many different disciplinary backgrounds immediately found their projects to fall within the boundaries of social capital research.

In a simplified manner, Putnam's central argument can be summarized as suggesting that our social environment exerts an enormous influence on the ways in which other components of our social system work. The formal and institutional-structural arrangements, according to him, can go only so far in regulating our social lives; there will remain, therefore, many areas that need to be taken care of by means of informal routines and inter-personal interactions. For instance, in the absence of institutions dealing with elderly, sick, and young children, family can step in; in the absence of an effective and inclusive welfare system, charity donations may take over; in the absence of a perfect market allowing free circulation of information and labour, social networks become instrumental.

The essence of what Putnam proposed, as Portes (1998) suggests, was not new. Similar ideas were put forward as far back as the late 19th century by, among others, Emile Durkheim. Discussing the forces that make a contract enforceable, Durkheim ([1957]) rejected the idea that the binding force of a contract came from the mutual obligations of the parties involved, because such a force is present even in the case of "unilateral contracts":

> [H]ow can the promise made by the other contracting party to fulfil certain terms of performance … compel me to honour my promise and vice versa? …Moreover, in order that there shall be a contract, there is no need for an undertaking of reciprocal performance. There may also be unilateral contracts… If, in a case of this kind, I declare that I will give a certain sum or some objects to a certain given person, I am bound to carry out my promise although I have received nothing in exchange. … How does it come to have this particular force? (p: 184)

In order for a contract to find a binding force, Durkheim argues, it needs to be somehow associated with a social force, or else the legal contract would not have any power to carry itself out: "[T]here is nothing in the word to bind the individual pronouncing it. The binding force, the action, are supplied from without. It is religious beliefs that brought about the synthesis…" (p: 194).

In an even more general way, Durkheim (1984[1893]) put law and morality parallel to each other, and explained that both were instrumental in enforcing a contract:

> The law of contract … exercises over us a regulatory action of the utmost importance, since it determines in advance what we should do and what we can demand ….[B]eyond this organized, precise pressure exerted by the law, there is another that arises from morals. In the way in which we conclude and carry out contracts, we are forced to conform to rules… There are professional obligations that are purely moral but that are nevertheless very strict. (p: 161–162)

Despite the old origins of the concept of social capital, it was certainly Putnam who gave the concept its recent visibility and relevance for the analysis of contemporary industrial societies. For Putnam (2000), social capital reveals itself through several dimensions (see Figure 1.1), each of which mediates between individuals and an aspect of their surrounding social environment. *Civic engagement* refers to the involvement of individuals in community affairs by virtue of membership in voluntary associations, from neighbourhood through national level. *Political engagement* shows the degree to which an individual is active in political affairs, whether it is membership in a political party or campaigning for one. *Religious engagement* looks at membership in a church and participation in activities organized by one. *Workplace engagement* refers to membership in trade unions and professional associations. *Volunteering* and *donation* highlight the degree to which one is prepared to sacrifice his or her time and money for the purpose of others' betterment. *Informal connections* point to the frequency of involvement in socialization activities. Finally, *trust and reciprocity* reveal one's underlying beliefs and feelings towards other members of his or her community. Putnam argues that these dimensions go hand-in-hand in creating a caring social environment wherein everyone is fully sensitive towards his or her community and also reasonably engaged in many, if not all, of the above activities.

Figure 1.1: Dimensions of Social Capital

The part of Putnam's argument that proved to be the most controversial and provocative involved his account of the changes in the levels of social capital over time. He based this argument on his study of two societies, Italy and the US. In the case of the former, Putnam (1993) argued that compared to the provinces in southern Italy, those in the north have consistently shown a higher level of economic development, as well as a more democratic political system; and, more importantly, they have also shown higher levels of trust among residents as well as a higher degree of civic engagement. As to whether the direction of causation has been from economic and political to civic and social or the other way around, Putnam argues that the changes in the former have not resulted in changes in the latter, but that the opposite has been true: social engagement leads to economic and political gains. This, for Putnam, was an indication that trust and civic culture have their roots in the social fabric of society, and not in the economy or polity.

With this conviction in mind, Putnam tackled the issue of social capital in the United States, and found that in the last third of the 20th century, American society has been experiencing a consistent decline in many indicators of social capital, such as the level of trust among people, confidence in government and public institutions, sensitivity towards national and communal affairs, membership in civic organizations, participation in elections, and readiness to volunteer (Putnam, 2000). In his opinion these trends send an alarming signal indicating that the foundation of the American civil society is gradually eroding, resulting in a corresponding erosion of American economic competitiveness and the democratic nature of its political system. Putnam later reinforced his key argument by elaborating on the connection between social capital and a large number of socioeconomic features of the American states, including changes in crime rates, deterioration of health conditions, and decline of life satisfaction over time (Putnam, 2001).

Soon after the publication of these works, two questions arose: Is America alone in experiencing this sinking pool of social capital, or is this part of a global trend? And, in either case, what has caused such a decline? In addressing the first question, Putnam himself edited a volume, *Democracies in Flux* (Putnam, 2002), in which the state of social capital in Britain, France, Germany, Italy, Spain, Australia, Sweden, and Japan was examined and extensively discussed. According to these reports, which revealed a variety of trends, the social capital reservoirs have remained stable and strong in Britain, Germany, Sweden, Spain, Australia, and Japan, have declined in Italy and the United States, and have gone through a radical transformation from being state-reliant to civil-society-reliant in France. Along the same lines, Fukuyama (1995a; 1995b) attempted to draw a social capital world map. Questioning the validity of the traditional classification of countries based on their degree of industrialization and the extent to which the state intervenes in the economy, he suggested the alternative classification of low- versus high-trust countries; Taiwan, Hong Kong, Italy, and France in the former group, and Japan, the United States, and Germany in the latter. Based on this new classification, Fukuyama argued that both sides of the ongoing debate over the appropriate role of the state in the economy – that is, the traditional left and right, the neo-mercantilists and neoclassical economists – have missed the point that non-rational factors such as trust will be crucial in the success of modern societies in a global economy.

As for the factors behind the decline of social capital in American society, Putnam (2000) pointed to the impact of generational change, pressures of time and money, suburbanization, and television. His overall assessment of the relative contribution of each of these factors is worth quoting at some length:

> Let us sum up what we have learned about the factors that have contributed to the decline in civic engagement and social capital ....
> First, pressures of time and money, including the special pressures on two-career families, contributed measurably to the diminution of our social and community involvement during these years. My best guess is that no more than 10 percent of the total decline is attributable to that set of factors.
> Second, suburbanization, commuting, and sprawl also played a supporting role. Again, a reasonable estimate is that these factors together might account for perhaps an additional 10 percent of the problem.
> Third, the effect of electronic entertainment – above all, television – in privatizing our leisure time has been substantial. My rough estimate is that this factor might account for perhaps 25 percent of the decline.
> Fourth and most important, generational change – the slow, steady and ineluctable replacement of the long civic generation by their less involved children and grandchildren – has been a very powerful factor. ...this factor might account for perhaps half of the overall decline. (Putnam, 2000: 283)

Other studies have diverged from Putnam's view on the factors behind this decline. Wuthnow (2002), for instance, attributed the decline to a corresponding decline in the socioeconomic resources of American citizens. He showed, for example, that between 1974 and 1991 the values of the social capital index dropped most noticeably among the most marginalized groups, such as blacks, those with larger families, less education, and lower income. This is so, in his view, because "people need to feel entitled in order to take part, and they need to feel that their participation will make some difference" (Wuthnow, 2002: 101); also, "people need other resources in order to create social capital, not least of which are adequate incomes, sufficient safety to venture out of their homes, and such amenities as child care and transportation" (p: 101). Along the same line, Theda Skocpol (2002) attributed the decline to the dampening effect of certain public policies. Focusing on the factors that led to the formation of a civic generation, she highlighted many specific historical events – such as war and revolution – and some specific policies – e.g., the formation of postal networks, the facilitation of the formation of local associations – that were conducive to a high level of civic engagement in an earlier golden era, and the lack thereof in recent decades.

There were also those who disagreed with Putnam that such a decline has really happened at all. Norris (2002), for instance, argued that while the decline in certain associational engagement is real, it has been compensated for by Americans' involvement in a host of new civic activities, such as participation in protest politics (e.g., petitions, demonstrations, and consumer boycotts), new social movements (e.g., environmentalism, feminism), and collective actions organized through the Internet. Wuthnow (1998), raises a similar point, arguing that the Americans are trying to come up with innovative ways to tailor their civic engagement to their new social realities – more mobility, more focus on work, smaller families, shorter and shallower social relationships. These new and innovative ways consist mostly of "looser, more sporadic, *ad hoc* connections in place of the long-term membership in hierarchical organizations of the past" (p: 5). For these authors, the change in the American social capital is more of a transformation in kind, rather than a drop in amount.

The Challenge of Diversity

A theme emphasized by some critics of Putnam – that social capital should not and could not be treated as a panacea for all social ills – was soon discovered by Putnam when he encountered the uneasy coexistence of social capital and diversity. Given the rapid increase in the ethnic and cultural diversity of many industrial nations, due to recent immigration from developing countries, any promotion of social capital has to occur in a culturally diverse environment. But, as Putnam (2003) found, with the rise of diversity in American states, the level of social capital declined. This finding was reiterated through another study at the neighbourhood level (published later in Putnam, 2007) and led to a controversy in British newspapers (see, for instance, Lloyd, 2006a; 2006b; Ulph, 2006) about the difficulty of keeping both diversity and social capital. Several other studies also supported Putnam's main argument in America but also in Australia and Europe (see, Coffe and Geys, 2006; Letki, 2008 forthcoming; Alesina and Ferrara, 2000; 2002; Leigh, 2006a; 2006b).

In his discussion of the relationships between social capital and diversity, Putnam (2007) raises three key points: 1) increased diversity and immigration are essential and inevitable, and generally strengthen advanced nations; 2) in the short-term, diversity and immigration challenge community cohesion; and 3) longer-term, successful immigrant societies overcome these challenges by building a broader sense of "we." Putnam's third point links the issue of diversity and social capital to another important area of recent interest, that is, the research on identity, which Putnam considers as 'an important frontier for research on social capital' (Putnam, 2007: 159).

Developing a sense of 'we' in a multicultural environment is not an easy task, as it became obvious from the reactions to the Multiculturalism (MC) polices of immigrant-receiving countries at about the same time as Putnam was discovering the uneasy relationship of diversity and social capital. Multiculturalism was a series of policies adopted in many countries to accommodate and celebrate the new ethnic diversity that was being brought in by their new immigrants. However, the support for Multiculturalism started to fade as a result of the heavy involvement of second-generation immigrants in a series of terrorist and/or violent activities that swept across the industrial nations, beginning with the 9/11 attacks in the U.S. in 2001, and followed by the assassinations of the Dutch politician Pim Fortuyn in 2002 and the Dutch film director Theo van Gogh in 2004, the Madrid train bombing in 2004, the London subway bombing in 2005, the social unrest in France in 2005, and the arrest of 17 terrorism suspects in Canada in 2006. If nothing else, such events sent an alarming signal that Multiculturalism has not been successful in generating a unified sense of 'we'ness out of the cultural and ethnic diversity of those societies.

The first indication of doubts concerning the merits of Multiculturalism surfaced in 2004 with the publication of a controversial editorial, entitled "Too Diverse?", in the influential British magazine *Prospect*. In the article, David Goodhart (2004) argued that in contrast to the 1950s, today's British society is witnessing much more 'value diversity' and 'ethnic diversity' resulting in the erosion of a common culture and identity, the weakening of social solidarity, the decline of support for social programs and the welfare state, as well as heightened concerns for safety and security. Predictably, the argument generated a flurry of counter-arguments, all published in the next issue of the magazine, by scholars from many different disciplinary backgrounds. A second sign surfaced in 2006 during a meeting of G8 countries in Lisbon. In their final report, the participant expert representatives of these countries acknowledged that: "[S]ome countries are reflecting on how to manage multiculturalism, while others are reflecting on whether or not to accept it as a model at all (Citizenship and Immigration Canada, 2006:18)." This sent a clear signal that the issue was more than just a British concern; it also declared the quiet arrival of what Ley (2005) has called, a 'Post-Multiculturalist' era.

The seriousness of this returning tide in response to MC can be felt everywhere. Recently, for example, one of the most optimistic advocates of MC in the 1990s, the Canadian political philosopher Will Kymlicka (2005) argued: "I confidently predicted that the overall trend across the West was clearly toward liberal multiculturalism… Today, however, it is clear that this prediction was false" (p: 82). There have also been, over a relatively short period, a number of publications and gatherings concerning this issue. In particular, the UNESCO launching of the interdisciplinary publication *International Journal on Multicultural Societies* clearly illustrates the attention this issue is receiving. Contributions to this journal have gradually shifted from a concern with religious and linguistic diversity, mostly in post-Soviet Eastern Europe, to an increased attention to the problems created by ethnic diversity in immigrant-receiving countries. The diversity proved to be nothing short of a serious challenge to many aspects of social life in the countries that were experiencing it.

Why Canada?

Canada is beginning to deal with the so-called challenge of diversity. Interestingly enough, Canada had a very slow start in social capital research but a very strong entrance into the area of investigating the relationship between social capital and diversity. The curious absence of Canada in the initial round of research on social capital can be seen through the small number of academic journal publications with a Canadian context since the early 1990s. As Figure 1.2 shows, between 1990 and 2008, the number of academic articles with a social capital theme rose sharply, from slightly more than 100 to about 2000; of these, those with a Canadian theme numbered only 5 in 1990 and about 50 in 2008. This gap shows that Canadian research on social capital is visibly lagging behind global trends. A symbolic illustration of this gap can also be seen in the curious absence of Canada in a very important volume edited by Putnam (2002), in which the state of social capital was discussed for several industrial nations other than Canada – i.e., The United States, Britain, France, Germany, Spain, Italy, Sweden, Australia, and Japan. Despite its lag in the general research on social capital, Canada was among the first countries to pay attention to the uneasy relationship between social capital and diversity, through a conference organized by Canadian Policy Research Initiative (PRI, 2003), at which Putnam reported the findings of his first study on the issue. This theme was later picked up by other Canadian scholars (see, for instance, contributions to the volume edited by Kay and Johnston, 2007).

Figure 1.2: Number of articles on social capital published in academic journals, 1990–2008

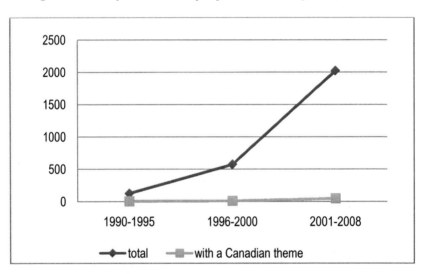

More Canadian studies on social capital, therefore, are needed not only for the benefit of those interested in Canada but also for the benefit of those concerned with other societies. As the American scholar, Seymour Martin Lipset (1997: 17) once said: "[T]hose who know only one country know no country;" that is, too narrow a focus on one country prevents observers from developing a comparative perspective, which could potentially draw their attention to many hidden corners of all societies being compared. The enrichment potential of comparative studies is greatly enhanced when an anomalous case is added to the mix; and, as far as the social capital dynamics is concerned, Canada might prove to be such a case, according to some existing signs.

Two such signs would suffice to be mentioned in this introductory chapter. The first, interestingly enough, comes from Putnam's (2001) study of the distribution of social capital in the United States. In this study, he breaks the comforting news to Canadians that, as one moves from southern American states towards states neighbouring the Canadian border, the stock of social capital increases systematically. Of course, the use of only American data stops him right at the Canadian border, but that task is taken on by the Canadian author, John Helliwell (1996). In a study entitled "Do Borders Matter?" Helliwell finds that the trend found by Putnam generally crosses over the border creating a typically higher level of social capital in Canadian provinces than in most American states. This higher level of social capital in Canada exists, despite the presence of many of the factors that Putnam identifies as the culprits behind the decline of social capital in the United States, – private entertainment, suburban life, and generational transformation.

A second sign for the presence of a potentially different Canadian story surfaced during the outpouring of rage by Muslim minorities in many western countries protesting against the publication of the cartoons of Prophet Muhammad in a Danish magazine. Surprisingly enough, such massive demonstrations did not appear in Canada. In search for the reasons behind this so-called Canadian anomaly, in an article titled, "Why global rage hasn't engulfed Canada," published in the Canadian daily newspaper *The Globe and Mail*, Valpy (2006) pointed to the possibility that Canadian multiculturalism and media might have played a role. Regardless of the causative factors, however, the mere presence of such unique Canadian dynamics point to the possibility of the existence of some sort of 'Canadian exceptionalism,' worthy of further investigation.

About this Book

I would like to conclude this introductory chapter with a couple of caveats and forewarnings. First, the goal of the present study has been, more than anything else, to draw an empirically-informed picture of social capital in Canada. This particular aim imposes its own limitations on the nature of the work, not the least important of which is the concise nature of the theoretical discussion of social capital in this volume. The full appreciation of the contents of this book, therefore, presupposes a certain level of familiarity with social capital literature on the part of the reader. Despite this, the theoretical discussion of the concepts makes it possible for readers with little prior exposure to this line of research to still follow the presented empirical information and to find them useful – or, so I hope.

Second, the difficulty of the measurement of social capital has often been mentioned in the literature. This problem is partly due to the multifaceted nature of the concept, and partly due to its newness. As a result, and despite the recent developments in the generation of social capital data, this research field still suffers from a lack of reliable data with all the necessary information. Therefore – and here lies the second caveat – like similar studies in many other countries, a comprehensive study of social capital in Canada requires the use of a vast array of data sources, from primary to secondary, and from quantitative to qualitative. This reality is reflected in the triangulated nature of the methodological approach adopted and the types of the data utilized in the present study. This diversity of methods and data types might make the readers who are used to only one type of data and one method more comfortable with certain chapters than others.

In this study, three main sources of data have been used. First, for the purpose of international comparison between Canada and other industrial nations, the World Values Survey (WVS) data have been utilized. The WVS consists of a series of surveys concerned with the worldwide political and socio-cultural trends, conducted in four waves in 1981, 1990–91, 1995–96, and finally in 1999–2001. More than 80 countries have participated in at least one of the above four waves, providing a rich set of data for international comparisons. Despite its great promise, the use of such data is not without problems. For one thing, each of the surveys included under the umbrella title of WVS has been conducted independently and, in most cases, through third parties. The reliability of the data, therefore, varies from one country to another and, for each country, from one wave of the survey to another. While this is a potential problem in general, the severity of it is less in the case of industrialized nations due to relatively standardized research approaches and a higher degree of transparency in research practices. The data are still fairly reliable as long as the aim is to capture general trends rather than to draw conclusions about finely-tuned and subtle affairs.

The second set of data come from a series of specifically Canadian surveys: the Canadian *General Social Survey* (cycles 1, 5, 11, and 17), *National Survey of Giving, Volunteering and Participating*, *Elections Canada*, *Political Support in Canada*, *Ethnic Diversity Survey*, and various *Canadian Census data* files: Profile, Individuals, Families, Households. In some cases, the data from various sources have been incorporated into one graph or table in order to examine the interactions between different variables. This practice is not without problems, given the different weighting schemes of different data sets; however, the relationships found can act, if nothing else, as sources of some hypotheses for future research using more systematic data.

The third source of data is based on the contents of a series of face-to-face interviews conducted during the years 2004–05. The main goal of the inclusion of such qualitative data was to delve into the deeper layers of the social capital experiences that survey instruments cannot easily capture. As it is obvious to social science researchers, the possibility of generalizing the findings derived from qualitative methods are limited, due to the small number of cases involved; however, the depth and richness that they bring to the understanding of the phenomenon under study make them indispensable for serious social scientific research.

The combination of the above multiple sources of information gives me the confidence – or, the hope, to be more precise – that at least the general traits of Canadian social capital have been adequately captured. Also, it is my hope that this volume will serve to identify the areas in which there exists a need for better and more sophisticated data, as well as the possible implications – both theoretical and practical – of studying social capital with an eye on diversity.

The chapters in this volume are grouped in four parts. Part One deals with the general social capital trends in Canada, both in isolation and in comparison to several other industrial nations. Part Two provides a big picture, discussing the multiple dimensions of social capital, and their profiles in different provinces, and among immigrants and the native-born Canadians. Part Three focuses on trust and its interaction with diversity. Part Four puts the discussions in the previous chapters together, by developing an integrated statistical model as well as adding the voices of the people who were interviewed. The next chapter starts this journey by comparing the Canadian state of social capital with that of nine other industrial countries.

Part I:

The Big Picture

2 Canada on the Global Map

As any researcher in the area of international development would immediately testify, an cross-national comparison of any phenomenon is a complex and daunting task. This, in most cases, has to do with the lack of comparable data. The recent availability of the World Values Survey data (WVS), however, has reduced the severity of this problem. The WVS consists of a series of surveys concerned with the worldwide political and socio-cultural trends and was conducted in four waves in 1981, 1990–91, 1995–96, and finally in 1999–2001. More than 80 countries have participated in at least one of the above four waves, providing a rich set of data for international comparison and producing some very interesting perspectives on global social and cultural trends (see, among others, Inglehart, 1999; Inglehart and Norris, 2003; Norris and Inglehart, 2004; and Norris, 2002). Canada was included in the first two and also in the fourth wave, allowing for some very interesting comparisons between Canada and a wide range of countries included in those three surveys.

The choice of countries to compare with Canada was made on the basis of the nations discussed in Putnam's (2002) edited volume, *Democracies in Flux*, which included at least a chapter on each of nine industrial nations: the United States, Britain, France, Australia, Italy, Spain, Sweden, Japan, and Germany, but none on Canada. Through the following pages the reader should be able to get a sense of not only where Canada stands in the world with regard to its social capital, but also the extent to which changes in Canada are similar to, or different from, those of other major industrial nations.

The Trends

Figure 2.1 illustrates the changes in the proportion of people in each country who discuss politics frequently. The highest level is reported for Germany, followed initially by Canada. However, the profile of the curves for both countries is an inverse U-shape, showing an increase in the proportion of those discussing politics during the 1980s and an almost equal decline throughout the 1990s. Except for Britain, the rest of the countries reported show a linear trend, either modestly rising or modestly declining (Spain). Despite the similarity of the shape of the curves for Canada, Germany and, to a lesser extent, Britain, the magnitude of increase and the subsequent drop for Canada is the most noticeable.

Figure 2.1: Discussing Politics Frequently: 1981–1999

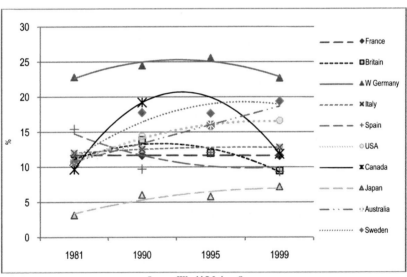

Source: World Values Survey

The data for a closely-related variable, interest in politics (see Figure 2.2), also reveals similar patterns. Here again, Germany reports the highest proportion, followed by Canada, and both countries along with Britain follow a curve peaking in the early 1990s. While the amount of change for Britain is relatively small, for Canada and Germany the changes are quite significant. This peculiar similarity definitely calls for an explanation, but let us see the rest of the data before we try to entertain any explanatory accounts.

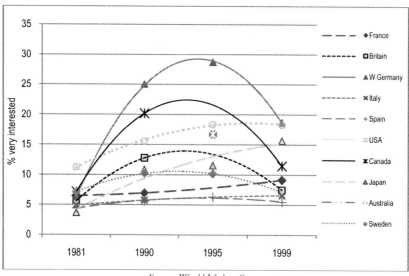

Figure 2.2: Interest in Politics, 1981–1999

Source: World Values Survey

Canada and Germany keep company in yet another area, that is, the proportion of the population who have signed petitions (see Figure 2.3). Again, the same curvilinear trends appear, making those two countries distinct from the rest of the nations compared here. The only difference in this case is that the percentages reported for Canada exceed those reported for Germany. The proportion of Canadians who have signed petitions also exceeds that of any other country at 1990, right before the beginning of the big downward turn that pushes Canada to the fifth place by the end of the decade.

Figure 2.3: Signing Petitions; 1981–1999

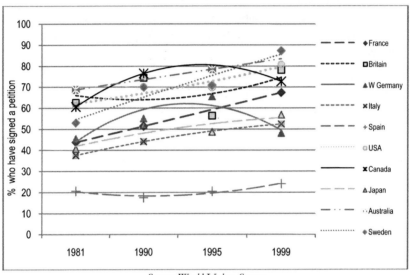

Source: World Values Survey

Figure 2.4 illustrates the changes in the proportion of population in each country who have participated in lawful demonstration. While France and Sweden show the highest levels, and the sharpest and most consistent increases over the whole period, Canada follows its own trademark pattern of rising in the 1980s and declining in the 1990s. Here, Germany follows Canada only half of the way and shows some stability in the rate percentages reported during the latter decade. In terms of the similarity of trends, Canada finds new company: Italy.

The trends of religious attendance shown in Figure 2.5 once again put Canada and Germany back in the same group, this time steadily declining over the whole period. Except for Spain, the numbers reported for all other countries show either no change or slight increases in the percentage of people who attend church regularly.

Figure 2.4: Participation in Lawful Demonstrations, 1981–1999

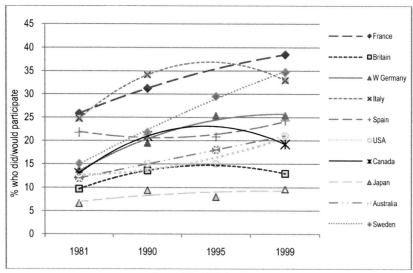

Source: *World Values Survey*

Figure 2.5: Church Attendance, 1981–1995

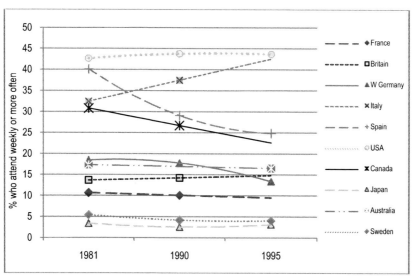

Source: *World Values Survey*

The trends for a closely-related variable, confidence in church, show a strikingly stable shape for almost all countries reported here. Except for Italy, as Figure 2.6 illustrates, people in the rest of industrial nations including Canada have consistently lost their confidence in the church. This is not surprising, given the seemingly global phenomenon of declining interest in organized religion (Norris and Inglehart, 2004).

Figure 2.6: Confidence in Church, 1981–1999

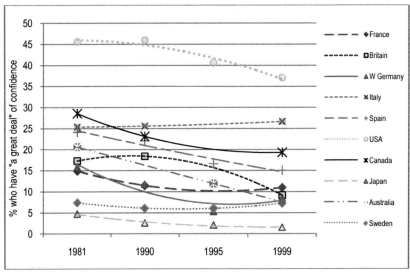

Source: World Values Survey

Data on the proportion of Canadians who consider their fellow citizens to be trustworthy follow a pattern similar to what we have seen before, that is, levels rise through the 1980s and decline in the 1990s. Figure 2.7 shows this pattern is mirrored by Italy. Of particular interest are the levels of trust reported for Sweden, reinforcing its reputation as a country with a consistently high level of social capital. The diagonal opposite of Sweden is France, with its consistently lowest rates of trust all throughout the two decades.

Figure 2.7: Trust in People, 1981–1995

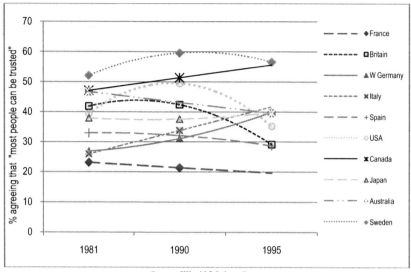

Source: World Values Survey

Like confidence in the church, the level of confidence expressed in the justice system also shows a relatively stable trend among all the countries that were compared, except for Japan (see Figure 2.8). Here again, all the trends are downward throughout the whole period. The magnitude of drop in the confidence in the justice system varies, of course, from one country to another, but Canada is among those with the sharpest drop over the last two decades of the 20th century. A more or less similar trend can be found for the reported level of confidence in the press (Figure 2.9).

The level of support that people are willing to provide to the nation in case of a national war is shown in Figure 2.10. Most of the curves move downward, but that of Canada, again along with Germany, moves upward slightly. Here again, Sweden shows the highest rate and the sharpest increase over the studied period.

33

Figure 2.8: Confidence in Justice System, 1981–1999

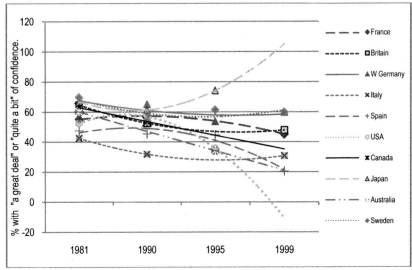

Source: World Values Survey

Figure 2.9: Confidence in Press, 1981–1999

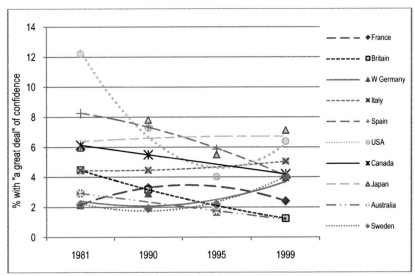

Source: World Values Survey

Figure 2.10: Willingness to Fight in a War for One's Country, 1981–1999

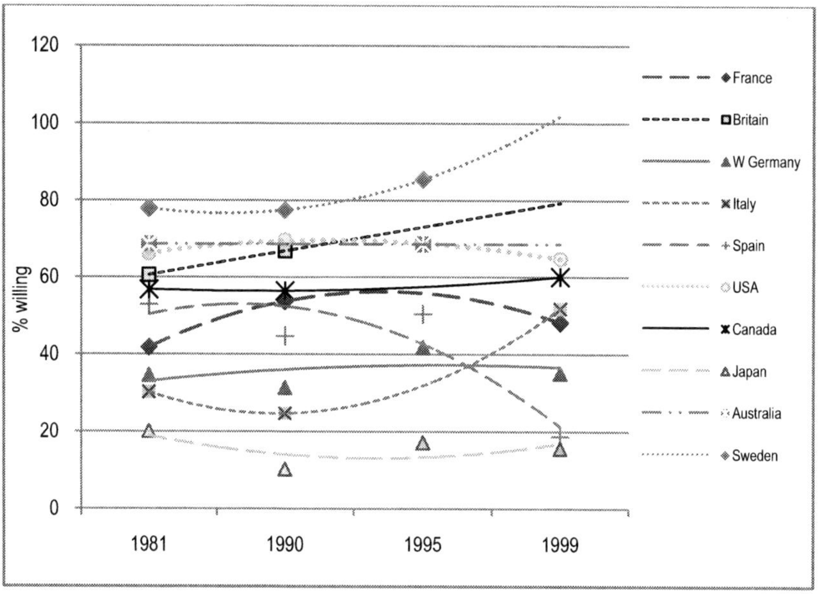

Source: World Values Survey

Conclusion

What do the above trends suggest? They seem to point to three major and relatively stable features of Canada's social capital. First, as far as the level of social capital is concerned, Canada positions herself somewhere in the middle of the continuum, at least compared to the countries included in the analysis. Second, as for changes in the stock of social capital over time, Canada shows a relative stability in most cases, although the early years of the 1990s seem to be acting as a watershed, signifying the beginning of a declining trend for many of Canada's social capital indicators. Third, the Canadian trends show a strikingly close resemblance to those of Germany.

The simultaneous examination of all the above trends leaves us with a mixed image. The relatively consistent increase of many indicators up to around 1990 is certainly an unusual thing for Canada, particularly given the fact that most of the indicators are related to politics, one way or another. The existing data do not tell us whether the normal state of affairs for social capital in Canada resembles that of the early 1990s when it appears higher, or the periods immediatly before or after that. The coincidence of the peak of the indicators of political engagement with the relatively heated political environment of Canada over the Meech Lake Accord raises the possibility that this increase was only temporary and due to the unusually over-politicized environment of Canada in those years.

However, even ignoring the raised levels of those trends around 1990, the values reported for the years before and after the increase place the country at the middle of the global social capital continuum. If we add to this the fact that some of the other indicators – such as the level of trust in others – witnessed stability or even an increase over the same period, the result for Canada's social capital report card would be even more distinct from the decline that characterizes many other industrial countries.

The reliance on WVS data, as mentioned earlier, makes cross-national comparisons possible, however, it limits the timeframe and the variables we are able to examine. In the next chapter we will focus only on Canada, getting a deeper picture by adding the information from several other Canadian data sources. This will allow us to capture the nature of social capital trends in Canada more accurately.

3 The Peaks and Valleys: Canada's Social Capital over Time

The previous chapter gave us a hint about where Canada stands on the global social capital map. Given the data points used, we could also get a sense of how things have changed during the period covered by WVS data. The main theme of this chapter is to move beyond international comparisons and focus specifically on Canada, in order to better understand the nature of changes in Canada's social capital over time. This exclusive focus on Canada allows us to utilize some additional sources of Canadian data, such as the *Canadian General Social Survey* (GSS), the *Political Support in Canada* (PSC) survey, and the *National Survey of Giving, Volunteering and Participating* (NSGVP). The inclusion of these data covers a longer period, hence a more reliable picture.

Despite the longer time range the above-mentioned data cover, its heaviest focus is again on the 1980s and 1990s. This poses some difficulty for comparability of the Canadian data with those of the U.S. that Putnam utilizes in his work; indeed, the 1980s are when most of the trends reported by Putnam actually end. This seems to be a problem we just need to learn to live with, at least for the foreseeable future.

The Peaks and Valleys

Figure 3.1 shows the changes in the proportion of Canadians who have reported affiliation with political parties and groups. It indicates that after a modest increase in the proportion of Canadians who claim membership in political parties from 1980 to 1990, the rates have started a consistent decline, to the extent that the percentage reported for 2003 is almost half that of 1990. Although the overall percentages for party membership and also the amount of change in those percentages are too small to yield a reliable conclusion, the nature of changes are consistent with those of several other indicators to be reported shortly, in that they all show a rise in the early 1990s and a decline after that. This indicates that there must have been a radical change in Canada's social environment since the early 1990s that has had a more or less dampening effect on various social capital indicators.

Figure 3.1: Membership in political parties/groups: 1981–2003

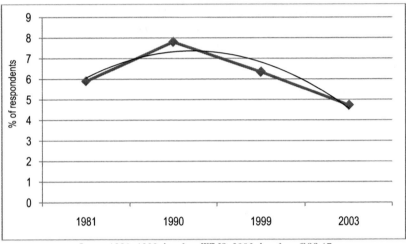

Source: 1981–1999 data from WVS; 2003 data from GSS 17

Over the same period, not only have the overall membership rates declined, but also has the nature of engagement of those affiliated with political parties gone through a transformation, that is, an increasing number of those affiliated with parties are becoming nominal and inactive members (see Figures 3.2 and 3.3). Such members would rather limit their support of their political parties to making financial contributions and possibly attending public meetings, rather than posting signs and getting themselves exposed in promoting the cause. Here again, the trend for active membership follows the exact same pattern as that of overall membership, that is, a peak around 1990 and a consistent decline ever since. With a small variation, the number of those attending political meetings follows similar patterns, that is, after 1990 there is a noticeable increase in the number of people who have never participated in such activities.

Figure 3.2: Active and inactive membership in political parties/groups: 1981–2003*

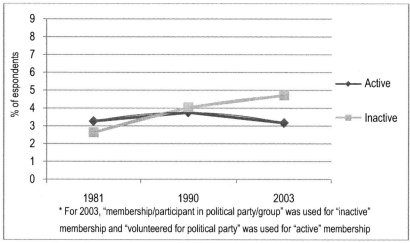

Source: Data for 1981–1999: World Values Survey; data for 2003: GSS 17

Figure 3.3: Frequency of attending political meetings or rallies: 1984–1993*

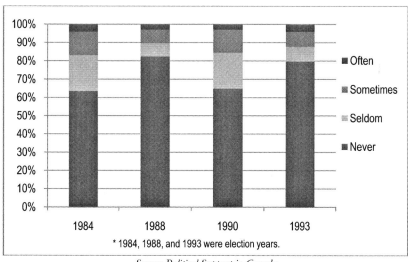

Source: Political Support in Canada

If Canadians are increasingly staying away from active involvement with political parties, does that mean that they are now embracing non-partisan politics in larger numbers? Figures 3.4 and 3.5 show that the proportion of Canadians who have signed petitions and/or participated in demonstrations or marches, as two prime examples of non-partisan politics, have experienced a similar decline over the period 1981–2003. Here again, the year 1990 acts like a watershed.

Figure 3.4: Signing petitions: 1981–2003*

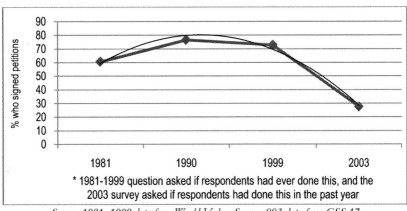

Source: 1981–1999 data from World Values Survey; 003 data from GSS 17

Figure 3.5: Participating in demonstrations or marches: 1981–2003*

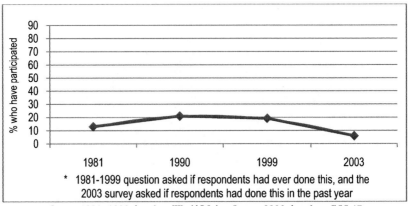

Source: 1981–1999 data from World Values Survey; 2003 data from GSS 17

A caveat to keep in mind while reading these two figures is the slight difference in the wording of the questions asked in 2003, in that the latter asks about any involvement *in the past year*. This specific focus on the previous year could dampen the number of affirmative responses, resulting in a more pronounced decline. But even without the last data point, the remaining trend indicates that the downtrend move had already started, resulting in a similar inverse-U shape.

The combination of the above two sets of trends point to a possibility that Canadians are becoming disillusioned with politics in general. Figures 3.6 through 3.8 provide a tentative confirmation for this hypothesis. The graphs show that the proportion of people who have expressed interest in politics and have frequently discussed political matters rose between 1981 and 1990, and then declined since 1990. The latter part of this trend is also reiterated by the drop since 1990 in the percentage of those who consider politics to be important.

Figure 3.6: Frequency of discussing politics: 1981–1999

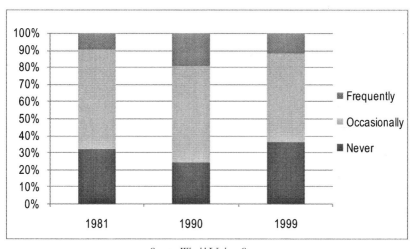

Source: World Values Survey

41

Figure 3.7: Interest in politics: 1981–1999*

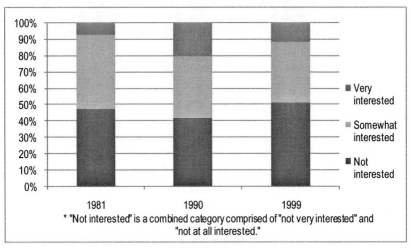

Source: *World Values Survey*

Figure 3.8: Importance of politics: 1990–1999

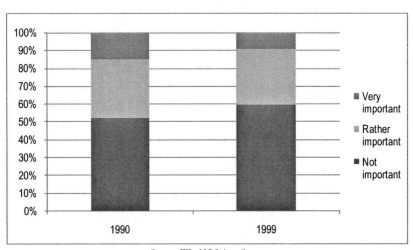

Source: *World Values Survey*

The above trends, which exemplify the declining trends in political participation, are also supported in large part by another more direct measure, that is, the proportion of people who have been showing up at the polls (see Figure 3.9). While the overall percentage remain relatively high for most of the period, a consistently declining trend emerges since the early 1990s, dropping from around 75% in 1988 to about 60% in 2004.

Figure 3.9: Voter turnout for federal elections: 1980–2004

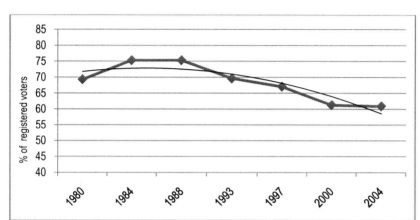

Source: Elections Canada

Figure 3.10: Joining unofficial strikes: 1981–1999

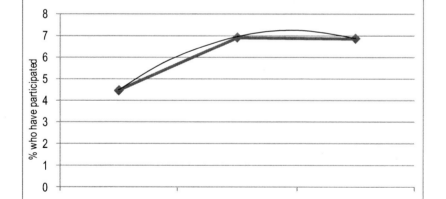

Source: World Values Survey

The lower level of voter turnout does not necessarily mean that Canadians are now involved in more militant, less peaceful activities, although a more or less similar trend there reinforces the trends we have already seen. Figures 3.10, 3.11 and 3.12 show that although there was an increase in the 1980s in the proportion of Canadians who joined unofficial strikes, participated in occupying buildings as protest and/or boycotted certain products, all such activities have either stagnated or declined since 1990. The small percentages of participants for all these activities, however, block further elaboration on this issue.

Figure 3.11: Occupying a building or factory: 1981–1999

% of respondents who have participated
3.5
3
2.5
2
1.5
1
0.5
0
1981
1990
1999

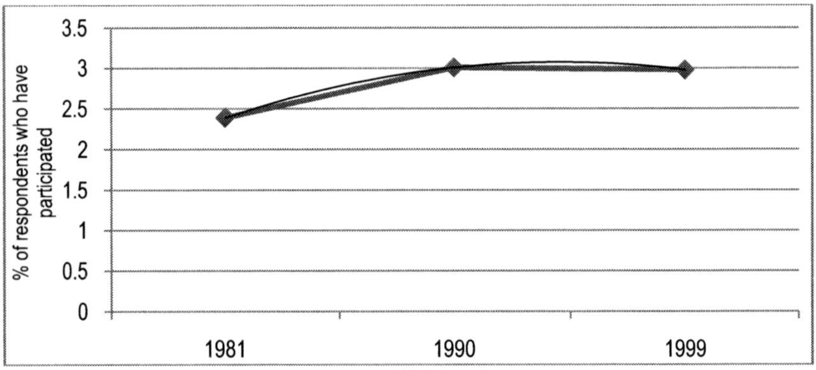

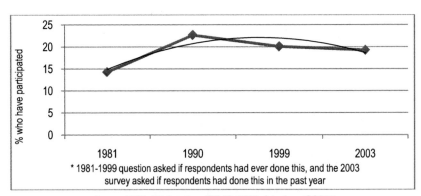

Source: World Values Survey

*Figure 3.12: Boycotting products: 1981–2003**

% who have participated
25
20
15
10
5
0
1981
1990
1999
2003
* 1981-1999 question asked if respondents had ever done this, and the 2003 survey asked if respondents had done this in the past year

Source: 1981–99 data from WVS; 2003 data from GSS17

Figure 3.13: Attitudes toward government: ("Most of the time we can trust people in federal government to do what's right"): 1984–1993

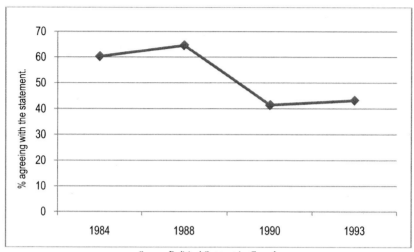

Source: Political Support in Canada

Figure 3.14: Confidence in government: 1981–1999

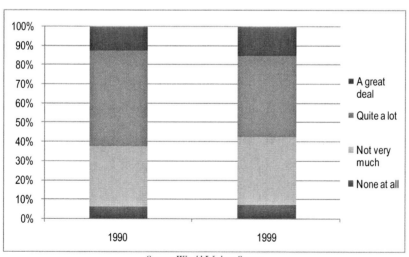

Source: World Values Survey

45

In trying to make sense of the above trends, one's attention would immediately be drawn to government as the main force alienating people from politics. Examination of the changes in level of confidence in various parts of government, however, does not fully support this argument. Figures 3.13 and 3.14 indicate that while the level of confidence in government had dropped during the 1980s, it remained constant or even increased during the 1990s, the same period in which the level of political engagement had started a downward trend.

Moreover, Figures 3.15 through 3.19 illustrate that, in particular, the level of confidence in parliament, the armed forces, the police, and civil services have either remained constant over this period or grown stronger. The only exception is confidence in the legal system, which declined between 1981 and 1990. Unfortunately we have no data from this source for possible changes after 1990.

Figure 3.15: Confidence in parliament: 1981–2003

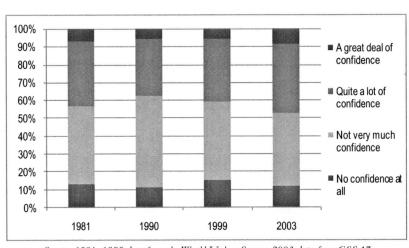

Source: 1981–1999 data from the World Values Survey; 2003 data from GSS 17

Figure 3.16: Confidence in the armed forces: 1981–1999

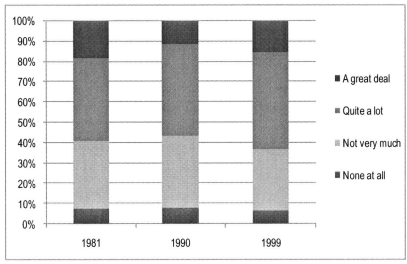

Source: World Values Survey

Figure 3.17: Confidence in the police: 1981–2003

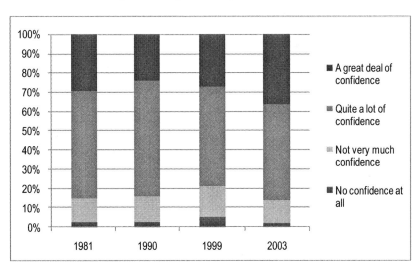

Source: 1981–1999 data from WVS; 2003 data from GSS 17

Figure 3.18: Confidence in the legal system: 1981–1990

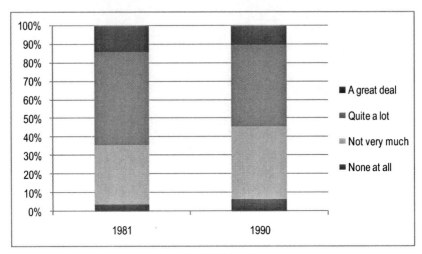

Source: World Values Survey

Figure 3.19: Confidence in the civil services: 1981–1999

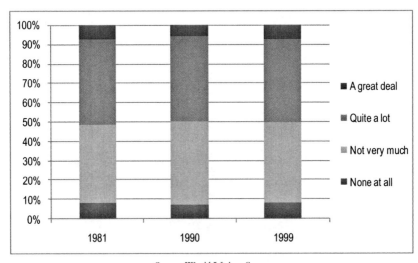

Source: World Values Survey

On a different front, engagement in church activities shows a seemingly paradoxical situation. For one thing, confidence in the church has been consistently dropping in the last two decades of the 20th century (see Figure 3.20). This is perhaps one of the most consistent trends out of those examined here, which, unlike the trends described above, shows no turning point in or around 1990. On the other hand, the trend for a closely related variable – i.e., the frequency of attending religious events (see Figure 3.21) – shows an opposite trend, as the proportion of those attending such events regularly remains relatively stable, and declines for those who never or rarely attend. In other words, the declining confidence in the church is combined with a relatively stable rate of church attendance. This might suggest a decline in the significance of church as a provider of belief systems and generator of meaning, but stability in its effect as a reinforcer of communal bonds.

Involvement in organizations such as sports clubs, service clubs, and political organizations is another aspect of social capital, for which, unfortunately, there is only limited data of a longitudinal nature. Figure 3.22 shows that between 1997 and 2000, membership in all the above organizations has been declining. Only two types of associations experienced growth over the period indicated: work-related organizations and cultural or hobby groups. Both of these groups are notable for being fairly inwardly-focused. Here again, the changes are so small that these results should be approached cautiously in the absence of any other supporting evidence.

Figure 3.20: Confidence in the church: 1981–1999

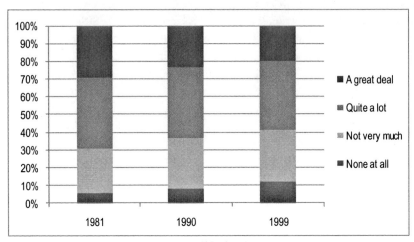

Source: World Values Survey

Figure 3.21: Frequency of religious attendance: 1981–2003*

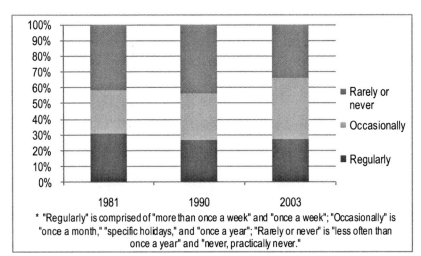

Source: 1981 and 1990 data from WVS; 2003 data from GSS17

Figure 3.22: Organizational membership of Canadians: 1997–2000

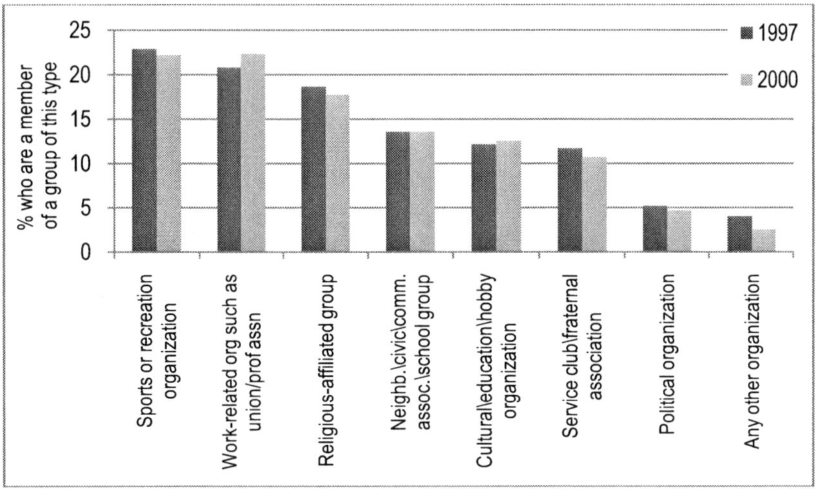

[Bar chart showing % who are a member of a group of this type, comparing 1997 and 2000, for categories: Sports or recreation organization; Work-related org such as union/prof assn; Religious-affiliated group; Neighb.\civic\comm. assoc.\school group; Cultural\education\hobby organization; Service club\fraternal association; Political organization; Any other organization]

Source: National Survey of Giving, Volunteering and Participating.

One implication of the data presented in Figure 3.22 may be that the work-related or work-based ties are rising in importance. Figure 3.23 on the donation methods of Canadians offers some support for this idea, as both donations through automatic payroll deduction and donations when asked by a co-worker have increased. Another consistent trend corroborated in this graph is attitudes toward religious institutions; as can be expected, lower confidence in the church seems to have depressed donations through that medium. In general, we can see that donations through friends and family have increased and donations through anonymous canvassers have decreased, what one may consider a personalization of philanthropy.

Voluntarism is closely related to donation; one is the sacrifice of time, the other money. The values illustrated in Figure 3.24 indicate that while the percent of people who reported volunteering dropped during the beginning of the 1990s, (a trend consistent with those reported earlier,) it has been growing ever since.

Figure 3.23: Donation methods: 1997–2000

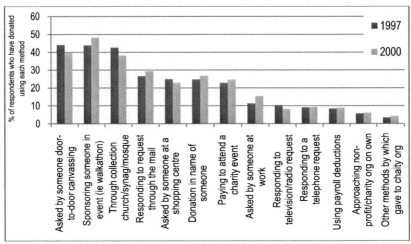

Source: *National Survey of Giving, Volunteering, and Participating*

Figure 3.24: Proportion of respondents that have reported doing volunteer work: 1990–2003

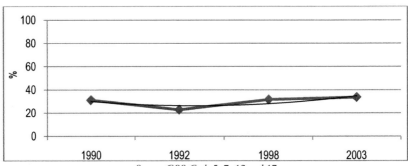

Source: *GSS Cycle 5, 7, 12 and 17*

The ways in which Canadians volunteer are also changing. Figure 3.25 shows that most forms of informal volunteering have declined, most notably, the four most common forms: shopping or driving for someone, providing unpaid babysitting, visiting the elderly, and writing letters or solving problems. Although yard or maintenance work, housework, providing care to the sick or elderly, and unpaid teaching or coaching have increased, their gains have been slight.

Figure 3.25: Informal volunteering activities: 1997–2000

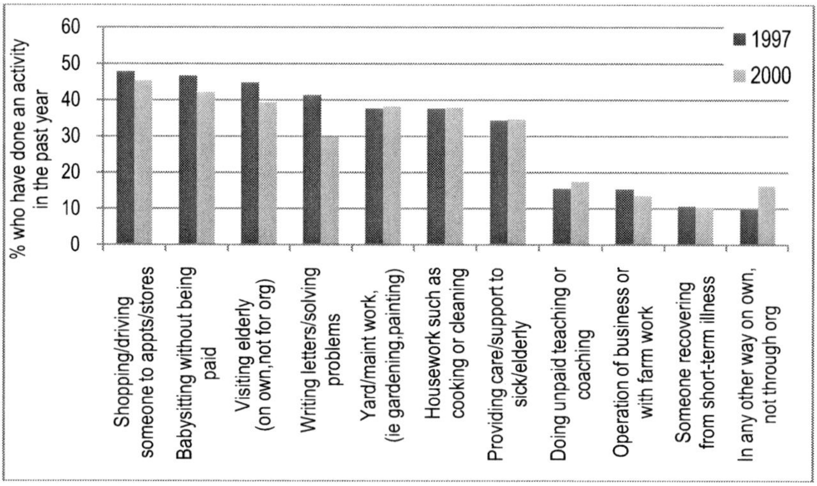

[Bar chart showing % who have done an activity in the past year, comparing 1997 and 2000, for activities: Shopping/driving someone to appts/stores; Babysitting without being paid; Visiting elderly (on own, not for org); Writing letters/solving problems; Yard/maint work (ie gardening, painting); Housework such as cooking or cleaning; Providing care/support to sick/elderly; Doing unpaid teaching or coaching; Operation of business or with farm work; Someone recovering from short-term illness; In any other way on own, not through org]

Source: National Survey of Giving, Volunteering, and Participating

One final trend in volunteerism is the noteworthy increase of employer support for the volunteer activities of their employees (Figure 3.26). This support takes many forms, from the use of facilities, to time off and official recognition. This is significant, as work could prove to be an obstacle to volunteering, as it demands time, energy, and possibly even money. These considerable incentives to volunteer should boost volunteerism accordingly and this could therefore be the cause of the rise indicated in Figure 3.26.

Figure 3.26: Employer support of volunteering: 1997–2000

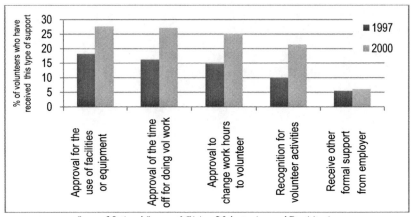

Source: National Survey of Giving, Volunteering, and Participating

Two other civil society institutions that are important in the creation and retention of social capital are labour unions and the press. Figures 3.27 and 3.28 show that the support for both of these follows the pattern we noted earlier: a relative stability in the 1980s and a declining trend in the 1990s.

Figure 3.27: Confidence in labour unions: 1981–1999

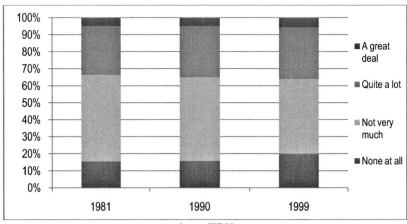

Source: WVS

Figure 3.28: Confidence in the press: 1981–1999

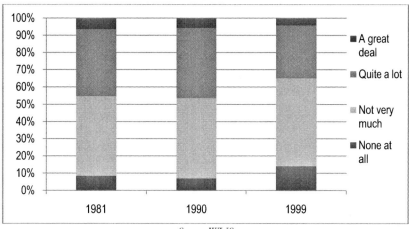

Source: WVS

One dimension of social capital in which Canadians have shown a noticeable increase involves attitudes towards their fellow citizens and their sense of belonging to the country as a whole. WVS data in Figure 3.29 show that the level of trust in other people has followed a similar pattern of increase in the 1980s and decrease in the 1990s; however, the data from GSS of 2003 show that the level of trust among Canadians is bouncing back to the level in 1990. A similar trend can be seen for the proportion of those who preferred having immigrants as their neighbours (see Figure 3.30). During the same period, there was a slight increase in the proportion of people who have expressed preparedness to fight in a national war or serve in the military if it becomes necessary (see Figures 3.31 and 3.32).

Figure 3.29: Trust in people: 1981–2003

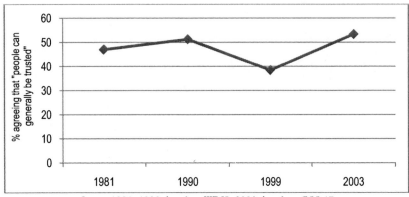

Source: 1981–1999 data from WVS; 2003 data from GSS 17

Informal social connections are also closely related to attitudes toward others. These connections are the day-to-day interactions we have with neighbours, friends, family, and even strangers. The information on how many friends a person has, how often they visit, and the leisure activities they do together are of significance here. In this respect, again, we are witnessing a similar trend, that is, a period of intensification of social capital in the 1980s and a decline afterwards. Figure 3.33 shows that between 1985 and 1990, there has been a noticeable drop in the propor-

Figure 3.30: Preference for having immigrants as neighbours: 1981–1999

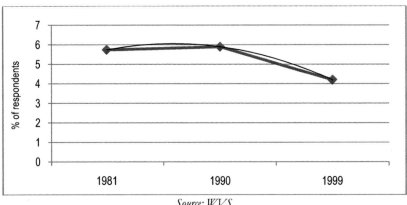

Source: WVS

Figure 3.31: Willingness to fight in war

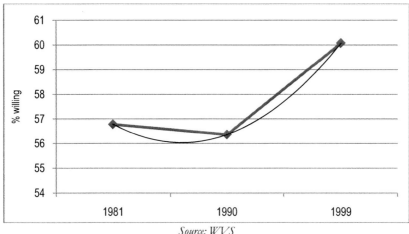

Source: WVS

tion of those reporting to have no or very few (one or two) close friends, and a corresponding increase in the number of those with a lot of friends (six through twenty or more). However, after 1990, both of these trends are reversed. In other words, for Canadians, the circle of friends is shrinking fast.

One might think that with a smaller circle of friends, there would be more time to spend with them. In other words, the loss of quantity of friends could be compensated for by a gain in the quality of friendship. Figures 3.34 and 3.35, however, show no obvious gain on that front, as people are contacting their friends less often than before; also, there are no significant changes in the frequency of people who contact friends via telephone very often.

Figure 3.32: Proportion of those agreeing that "People should be willing to serve in the armed forces if the government asks them, even if they don't want to": 1984–1993

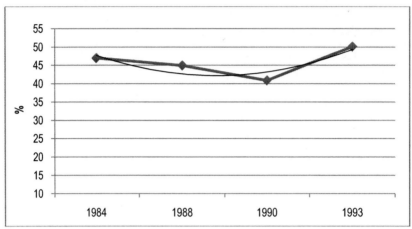

Source: *Political Support in Canada*

Figure 3.33: Number of people considered close friends by respondents: 1985–2003*

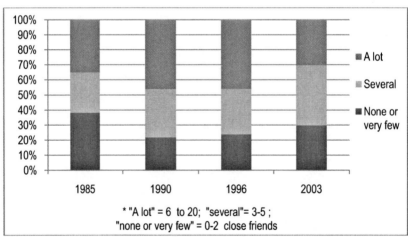

Source: *GSS Cycle 1, 5, 11, and 17*

Figure 3.34: Frequency of contact with friends: 1985–2003*

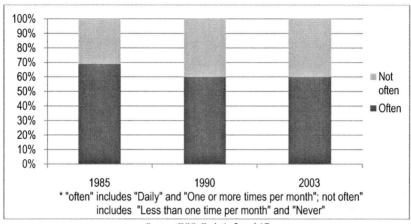

Source: GSS Cycle 1, 5 and 17

Figure 3.35: Frequency of contact with friends by telephone: 1985–2003

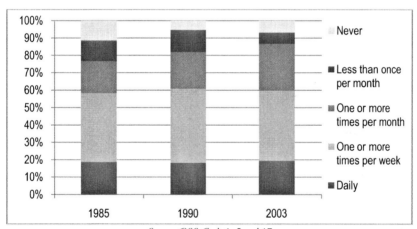

Source: GSS Cycle 1, 5 and 17

Finally, there is partial evidence for the presence of a similar trend in participation in sports. While the existing data do not cover the 1980s, those for the 1990s show a declining trend, consistent with those observed earlier (see Figure 3.36).

Figure 3.36: Regular participation in sports: 1992–2003

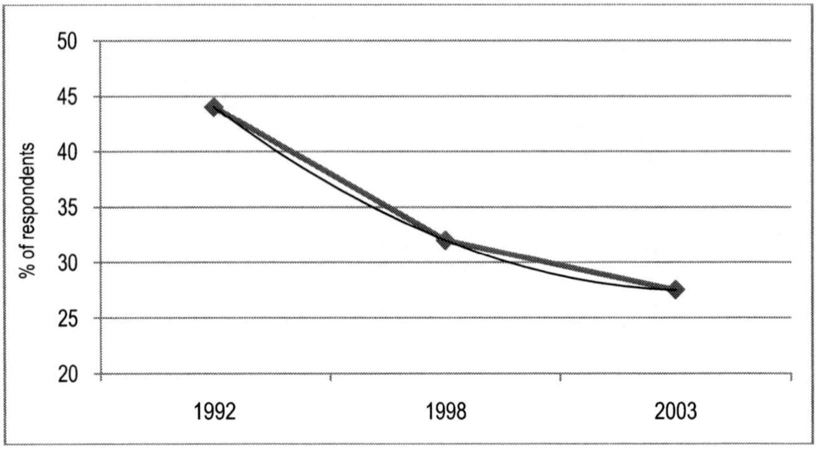

Source: GSS Cycle 7, 12 and 17

Conclusion

The existing Canadian data on various indicators of social capital do not allow for capturing the trends over an extended period of time. The three waves of WVS that are used in this study cover mostly the last two decades of the 20th century. As far as this period is concerned, it seems that Canada has not been experiencing the consistent decline of social capital similar to what Putnam and others have found for the United States since the 1960s. This is good news by and in itself.

However, despite the relative stability of the Canadian social capital reservoir, the data examined here show that there are two relatively distinct sub-periods within these two decades; during the 1980s, most social capital trends were either stable or on the rise; and since the early 1990s, they have been plummeting. The explanation of this change is of great significance to both Canadian policy-makers and social scientists. Before discussing some possible hypotheses in this regard, let us further refine the question to be answered a bit further. The above-mentioned decline of Canadian social capital since the early 1990s has not occurred for all social capital indicators. The indicators which experienced the most visible decline are: involvement with political parties, participation in political activities, the level of political sensitivity, confidence in church, and the size of personal networks of friends. On the other hand, there have been stability or modest gains over the same period in the areas of confidence in various governmental institutions, trust in other countrymen, volunteering, and willingness to fight for the country if necessary. The question then becomes: why such divergent trends?

One possibility, as Putnam argues in the case of the U.S., is the pressure of time and money. This may sound like a valid hypothesis, given the severe economic recession of the early 1990s which resulted in a rise in the national poverty rates. However, the areas in which the impact of an economic recession tends to be most visible are those involving people's willingness to sacrifice time (volunteering) or money (donation), and neither of these two shows a decline during the 1990s. Putnam has also pointed to TV and the amount of time people spend watching it as another culprit in killing the civic vitality in the United States. This does not seem to explain the shift of trends in Canada either, as the hours of watching TV has been constantly on the rise for both decades.

Do the Canadian trends of social capital have to do with the issue of the particular situation of Quebec within the country? One thing that points to the potential relevance of this possibility is the striking similarity that we have found, based on the analysis of WVS data, between Canadian trends and those of Germany, a country that was made by the merger of two previously separate nations at the end of the 1980s. As in the case of Canada, the early 1990s acted as a turning point for many of Germany's social capital trends. In both cases, it is possible for such duality in the cultural and ethnic fabric of the society to artificially raise the frictions between the two segments, which will in turn lower the degree of connectedness between them.

Despite the relevance of this possibility, this hypothesis cannot explain the particular nature of the Canadian social capital trends. First, such confrontations normally lower the level of trust among people, and that is exactly opposite to what happened in Canada during the 1990s, a period in which the separatist sentiments in Quebec noticeably intensified. Second, the encounter between a minority and majority typically results in an intensification of political involvement and sensitivity on both sides, and this again is opposite to what has happened in Canada in the 1990s.

Yet another hypothesis could be entertained here. Is it possible that the Canadian trends have been a product of the impacts on the Canadian psyche of the recognition of Canada by the United Nations as 'the best country to live in' during the 1990s? When the residents of a country believe and internalize such a perception, they may tend to become more forgiving towards their government as well as to their fellow citizens, as such recognition can be perceived to be a product of the ways in which the government has handled the nation's affairs. A high and relatively constant level of confidence in various parts of government, and willingness to participate in a national war to defend the country can be products of such a mindset. On the other hand, when people's perception is that the functioning of various institutions in society are reasonably effective, they are more likely to see no particular reason for their own heavy involvement in political affairs, which is what seemed to happen in Canada towards the latter half of the 1990s.

A true test of any of the above hypotheses requires the use of reliable longitudinal data with consistent measurement of various social capital indicators over a long period of time. Such data are, unfortunately, unavailable. In the absence of such data, our next best choice is to use cross-sectional data, in order to find the correlates of various indicators of social capital. One could, for instance, compare different provinces, cities, and neighbourhoods; alternatively, one could compare different groups, and examine the factors associated with their reported levels of social capital. In the next few chapters, we start a journey in that direction.

PART II:

Multi-Dimensionality of Social Capital and
Multi-Dimensionality of Diversity

4 A Close-up on the Concept: Dimensions of Social Capital

The current and predominant understanding of social capital is heavily influenced by the way in which Robert Putnam conceptualized it in his *Bowling Alone* (2000). In this conceptualization, there are two elements that have been taken for granted without any attempt to empirically verify them. One involves the various dimensions of social capital – trust, volunteering, informal networks, religious engagement, political participation, reciprocity, confidence in institutions, and donation – which Putnam uses to organize his discussion on American social capital trends. There is no serious discussion as to whether these are supposed to be treated as various reflections of what should be known as social capital or, alternatively, these are just some headings to be used for a better and more effective organization of the materials. The second, and perhaps more important, question here is this: assuming that these are different dimensions of social capital, do they always go hand in hand, or can they trigger conflicting forces that might cancel each other out? In other words, are we to assume an internal consistency among various dimensions of social capital, or we can treat them as independent forces?

The answers to the above two questions are of utmost importance, because of their theoretical and practical implications. If it appears that the suggested dimensions do not empirically exist, the concept of social capital seems to be reduced to a mere organizing principle to be used only for the purpose of improving the effectiveness of the presentation of some old information. On the other hand, if it is shown that these dimensions do really exist, but they do not always move hand in hand with each other, the policy-makers concerned with improving the stock of social capital would then need to focus on some dimensions at the expense of others. Lack of such conceptual clarity has resulted in a situation in which, according to Lin (1995), "[s]ocial capital has become too vast, including too many elements at different levels of analysis, to be empirically specified apart from human capital and economic resources" (quoted in Kay and Bernard, 2007: 42). Making such distinctions is now a timely endeavour, and the social capital research seems to have matured enough to benefit from such scrutiny.

Against this background, an attempt is made in this chapter to get a close-up image of social capital using the data from cycle 17 of the Canadian General Social Survey. This survey, which was conducted in 2003, is particularly social-capital-rich, as it contains the answers of more than 25,000 Canadians to more than 250 questions all related to the various aspects of their social engagements. The generation of such a rich set of data has, in a sense, put Canada in the same league with many other industrial nations that had already developed their own social capital surveys[1]. This practice has two potential benefits. First, as mentioned above, it will help us acquire a more accurate and empirically-based picture of the dimensions of social capital, a badly needed thing, given the still heavily theoretical nature of most of the existing debates in social capital literature. Second, it will 'Canadianize' the conceptual backdrop of the present study, which is a logical thing to do in a study concerned with Canadian realities.

A brief note on a couple of methodological points is in order here. I should admit, though, that to a general reader who is not familiar with multivariate statistics, the next couple of paragraphs might feel like a rough ride. That is always the case with methodological notes, as they tend to make things look and sound more abstract, more complicated, and less fun than they really are. I assure you, though, that after reading through the next couple of paragraphs, the rest of the chapter will feel much smoother.

Here is the first methodological point: to capture the structure of the relationships among the social capital variables, I have employed a statistical procedure called Principal Component Analysis (PCA). As a data reduction technique, PCA disentangles a complex web of relationships that might exist among a large number of variables, and reduces the initial complexity by placing the involved variables under a few headings,

---

1 In the United States the survey was called "The Social Capital Community Benchmark Survey," and in Great Britain, "The ONS Social Capital Project." This project has been underway since 2002 through a "cross-governmental social capital working group."

namely, the underlying 'components' (or what we have called here, the 'dimensions'). Each of the resultant dimensions encompasses the variables that are most similar or most closely related to one another. This results in a more concise, more manageable, and more easily interpretable picture.

The second methodological point is that, while I was initially interested in including all the variables discussed by Putnam, I was limited to those available in the Canadian GSS 17. This, however, was not a serious limitation, as it still left a long list of variables (45 of them, to be precise), all of which were relevant for the purpose of our study. These variables are listed in Table 4.1 below.

To avoid losing the reader in the midst of the detailed discussions to follow, here is a bird's eye view of the steps followed and the order in which the materials will be presented in the rest of this chapter. First, as an eligibility requirement for variables to be used in PCA, several non-interval variables have been converted into a series of dummy variables – that is, variables that take only the two responses of yes or no, represented by the values of 1 and 0, respectively. Also, based on the results of an initial run of the model, some of the variables have been dropped due to their weak correlations with the rest of the variables. Third, a PCA has been run, to see how the variables naturally sit together without having to be shoehorned into some pre-constructed dimensions. Fourth, using the results of this exploratory PCA, a series of *composite indexes* have been constructed, each representing one distinct dimension of social capital. Such indexes have then been used to report the social capital profiles of different regions and groups, the results of which will be reported in chapters five and six.

Table 4.1: The Social Capital Related Variables Included in the Principal Component Analysis

Past year: member/participant in union/professional association
Past year: member/participant in political party/group
Past year: member/participant in sports/rec organization
Past year: member/participant in cultural organization
Past year: member/participant in religious affiliated group
Past year: member/participant in school group/neighbourhood association
Past year: member/participant in service club/fraternal organization
Past year: member/participant in any other type of organization
How frequently participate in group activities and meetings
Past month: have you done a favour for a neighbour?
Past month: any neighbours done a favour for you?
Did you vote in the last federal election?
Did you vote in the last provincial election?
Did you vote in the last municipal or local election?
Past year: searched for information on political issue
Past year: volunteered for political party
Past year: expressed view by contacting newspaper/politician
Past year: signed a petition
Past year: spoke out at a public meeting
Past year: participated in a demonstration or march
How frequently do you follow news and current affairs?
Past month gave help: teaching, coaching, practical advice
Frequency of religious attendance of the respondent
Importance of religious/spiritual beliefs to how live life
How many other friends (neither relatives or close friends)
Past month: how often did you see your friends
In general, people can be trusted
How trustworthy: people in your family
How trustworthy: people in your neighbourhood
How trustworthy: people in your workplace or school
How trustworthy: strangers
Confidence in police
Confidence in judicial system
Confidence in health care system
Confidence in school system
Confidence in welfare
Confidence in government
Confidence in bank
Confidence in major corporation
Confidence in business people
Past year: did unpaid volunteer work for any organization
On average how many hours per month did you volunteer?
Past year: donate money/goods to organization or charity
While in grade/high school, particpated in organized team sport
While in grade/high school, belonged to a youth group

The Dimensions of Social Capital: The Results of the PCA Model

The results of the PCA model are presented in Table 4.2. As it is clear from the table, the 45 variables related to social capital have been clustered into 15 broader groupings, i.e., components or dimensions. Let's start with the first cluster of variables, that is, those variables that have the highest correlation with the first component. It is easily noticeable that all those variables have to do with trust: trust in people in general, in family members, neighbours, co-workers, and strangers. The values reported for each of those variables indicate the degree of association between that variable and the common force that is pushing all of them forward. Those values can vary from -1 to +1, where 0 signifies a total lack of correlation and an absolute value of 1 a perfect correlation. The negative or positive signs of those values show the direction of such associations; a negative value for a particular variable indicates that as other variables rise, that particular variable drops. The fact that all the values reported for the first cluster are positive indicates that all the trust variables go hand in hand with one another, i.e., if a person tends to find the general public trustworthy, he or she tends to also trust family members, neighbours, co-workers, and even strangers.

This small finding can have another, perhaps more theoretically important, implication. Some have argued that the true test of trusting behaviour is whether or not one can trust strangers. Offe (1999), for instance, argues that people always have a great deal of trust in their family members, neighbours, co-workers, and generally all those with whom they have a face-to-face relationships for a long period of time; what shows whether or not a society is indeed trusting is the extent to which people can trust those with whom they have no regular and face-to-face contact. The positive and relatively strong correlations between the trust variables in Table 4.2, however, shows that the tendency to trust has a spill-over effect; that is, it will extend from those with whom one has a primary relationship to those with secondary contacts, and vice versa. In this case, the trusting behaviour seems to be more of a general tendency, possibly shaped either by major events in life or through the primary socialization process, as opposed to being a fluctuating situational phenomenon that would constantly change in response to minor positive or negative experiences.

The fact that trust has surfaced as the first dimension indicates that this is one of the strongest and most coherent dimensions of all. In a more technical language, this means that, out of the 15 dimensions which surfaced, trust captures and explains a larger portion of the common variance that exists among the included variables. This sets trust aside as one of the major components of social capital, something reflected in the fact that a large body of literature has treated trust either as a main element, or as a surrogate, of social capital. Due to its significance, therefore, trust has been given a central stage in the discussions in the latter half of this book.

The second cluster has to do with *confidence in main institutions*: the education system, health care system, welfare system, government, judiciary, and police. The underlying commonality in all these institutions, at least in the Canadian context, is that they are all related to the government. Here again, the numbers reported for all variables are positive, meaning that confidence in each of the six institutions is positively correlated with confidence in the rest of them. This is to say that, despite the relative independence of different branches of government – judiciary, and executive – people see them as part of a whole package, and their confidence in each component of the package goes hand in hand with the rest.

The third principal component has to do with *voting*. The three variables clustered under this component are indicators of whether or not the respondents have voted in the last federal, provincial, and municipal elections. Again, the variables are positively correlated, and strongly so, indicating that the tendency to vote in elections does not vary much by the level of government in which the election is held. If someone cares to vote, they will vote regardless of the type of election, and if someone is not willing to participate in elections, they will not, again regardless of the type of election. Given this strong correlation, it seems that the tendency to vote is informed by a force that has nothing to do with the type of election, but perhaps more with the perception of whether or not the voting makes any difference. When the answer to this question is positive, then people vote in any election, and when the persuasion is not there, they will not vote regardless.

The fourth principal component has captured three variables that are all related to *religion*: the importance of religious/spiritual beliefs, the affiliation with a religious group, and the frequency of attending religious functions. The positive and strong correlations reported for all these variables are something that makes sense intuitively, that is, those who value spirituality and religion are more likely to be affiliated with a religious group and also to attend religious events.

The fifth principal component shows an interesting combination of variables. The essence of the component is about *volunteering*, i.e., whether or not one has done any volunteer work, and the amount of time devoted to such work; however, these two variables are also correlated with a third one, that is, membership in organizations other than the ones captured by other variables in the analysis (i.e., political, religious, sport, and cultural organizations, service clubs and fraternal organizations, labour unions, as well as neighbourhood associations). The inclusion of this variable in the volunteering component would mean that these are the types of organizations that need the volunteer energy of their membership or, alternatively, these organizations attract a certain group of people who are more willing to volunteer. Also, this type of volunteering can be considered a general-purpose volunteering, different from that which is done for specific purposes such as supporting a political party, to which we turn now.

The sixth principal component highlights *engagement with political parties*. This is different from general political engagement captured by other variables. Under this component, there are two variables with high and positive correlation scores: membership in political parties and volunteering for them. Obviously, membership in a political party signifies a high level of commitment to a cause and, hence, the sacrifice of time and energy to promote the cause is perfectly understandable.

Neighbourliness is probably the best description of what the seventh component illustrates. The two variables included under this component are whether or not one has done or received a favour from his/her neighbours. The strong and positive correlation scores are an indication that when one is present, the other is very likely also to be present. In other words, there is a reciprocal and mutual element in one's relationship with their neighbours.

Table 4.2: Factor Loadings for Social Capital Variables (Rotated Component Matrix)

	1	2	3
People can be trusted	0.864		
How trustworthy: people in your family (very)	0.486		
How trustworthy: people in your neighbourhood (very)	0.721		
How trustworthy: people in your workplace or school (very)	0.695		
How trustworthy: strangers (very)	0.752		
Confidence in police		0.378	
Confidence in judicial system		0.582	
Confidence in health care system		0.631	
Confidence in school system		0.635	
Confidence in welfare		0.640	
Confidence in government		0.621	
Voted in the last federal election			0.877
Voted in the last provincial election			0.881
Voted in the last municipal or local election			0.780
Past year: member/participant in religious affiliated group			
Frequency of religious attendance of the respondent			
Importance of religious/spiritual beliefs to how live life			
Past year: member/participant in any other type of organization			
Past year: did unpaid volunteer work for any organization			
Average number of hours of volunteering per month			
Past year: member/participant in political party/group			
Past year: volunteered for political party			
Past month: have done a favour for a neighbour			
Past month: any neighbours has done a favour for			
Past year: searched for information on political issue			
Past year: expressed view by contacting newspaper/politician			
Past year: spoke out at a public meeting			
Past month gave help: teaching, coaching, practical advice			
Confidence in bank			
Confidence in major corporation			
Past year: member/participant in sports/rec organization			
Frequency of participating in group activities and meetings			
While in grade/high school, participated in organized team sport			
Past year: signed a petition			
Past year: participated in a demonstration or march			
Number of friends (neither relatives or close friends)			
Past month: frequency of seeing friends			
Past year: member/participant in cultural organization			
Past year: member/participant in school group/neighbourhood association			
Confidence in business people			
Past year: donated money/goods to organization or charity			
While in grade/high school, belonged to a youth group			
Past year: member/participant in union/professional association			
Past year: member/participant in service club/fraternal organization			
Frequency of following news and current affairs			
Extraction Method: Principal Component Analysis. Rotation Method: Varimax with Kaiser Normalization.			

	Component										
4	5	6	7	8	9	10	11	12	13	14	15
0.781											
0.843											
0.632											
	0.334										
	0.822										
	0.815										
		0.862									
		0.870									
			0.879								
			0.868								
				0.574							
				0.567							
				0.581							
				0.357							
					0.728						
					0.738						
						0.742					
						0.502					
						0.570					
							0.676				
							0.726				
								0.566			
								0.722			
									0.519		
									0.671		
										0.387	
										0.462	
										0.656	
											0.455
											0.502
											0.536

The eighth principal component attracts four variables: searching for information on political issues, expressing views by contacting newspapers and politicians, speaking out at public meetings, and providing help by teaching, coaching, and giving practical advice. The central element in all these variables seems to be the act of *exchange of information, mostly on political and social issues*. This can include both the acquiring and the dissemination of information. While the last variable does not specifically talk about the political nature of the advice given and teaching done, it seems that the eagerness to acquire information on political issues has a spill-over effect into other areas of life which are less political in nature.

The ninth principal component is somewhat related to the second one as both deal with *confidence in institutions*, though with a major difference. While the former looks primarily at public institutions, the latter involves *private-sector* and for-profit organizations such as banks and major corporations. This distinction is quite interesting, as it indicates that these two sets of variables are not necessarily correlated with each other, and that each is capturing a different strand of the confidence phenomenon.

The tenth principal component incorporates membership and/or *participation in sports and recreational activities* both currently, and previously while at school. The positive correlation between these two indicates that involvement in these kinds of activities at a younger age has a lasting effect on the likelihood of similar engagement in later years. Also, both of these variables are correlated with the frequency with which people participate in group activities and meetings. This points, partially, to the significance of the socialization effect, that is, teamwork activities done during the primary socialization period make it easier and more probable for one to get involved in group activities later in life. This is a confirmation for the point Uslaner (1999) makes regarding the significance of sports clubs in generating and promoting social capital.

The eleventh principal component involves two variables: signing petitions and participating in demonstrations, both of which have to do with the expression of dissent over political and social issues, best described by the term proposed by Norris (2002): *protest politics*. One might find a lot of similarities between this and two other components – the sixth one on engagement with political parties, and the eighth one on the exchange of political information. A closer examination of the variables included under each of the three components, however, shows that while the sixth one is looking at activities organized through political parties and the eighth one includes activities revolving around the getting and giving of information, the eleventh component looks mostly at irregular political activities that are temporary in nature and do not create a commitment on the part of the doer to continue their engagement for a prolonged period. In other words, the latter involves mostly sporadic, as opposed to regular, engagement.

The essence of the twelfth component seems to be what Putnam calls *informal networks*, that is, socialization and time-spending with friends. The two variables included in this component are the number of friends one has, and the frequency with which they see their friends. It goes without saying that a larger social network calls for more time to be spent with friends. As mentioned before, such social networks do not have a very strict structure and agenda and are, instead, very fluid in nature. They can have an underlying political, religious or cultural basis, or they can be free-floating webs of relationships. In either case, both the presence of such networks and their sheer size highlights the involvement of individuals in social relationships with their fellow countrymen.

The thirteenth principal component demonstrates an unusual combination of variables: *participation in cultural activities* and *membership in school groups and/or neighbourhood associations*. From a certain angle, this sounds similar to the tenth component that showed the high correlation between involvement in sport clubs at school and participation in group meetings and activities. This clearly points to some sort of continuity between group involvement at school at a younger age and engagement in society at a later time. Furthermore, it seems that even the nature of the latter engagements is influenced by the former ones. Involvement in school team sports, for instance, leads to a similar involvement later on, both geared towards generating pleasure and entertainment for the doer; involvement in cultural activities at school is more strongly associated with involvement in civic organizations with more serious mandates.

Confidence in business people, donating money for charity purposes, and belonging to a youth group at school are heavily loaded on the fourteenth component. Membership in youth groups seems to be nurturing a caring for others and, hence, can lead to a higher level of generosity in form of donating money. Also, since the money donation seems to be more frequently done by people who have their own businesses – both because they can afford to do so, and also because it wins them tax rewards and good reputations – there is a possibility that a larger proportion of these individuals are themselves business people. The possible connection between the membership in youth groups and involvement in business – if the above proposition happens to be valid – remains an area open for further investigation.

The last principal component combines three variables: *membership in labour unions, participation in service club and fraternal organizations,* and *the frequency of following news and current affairs.* A comment element in the first two variables can be the self-centered nature of the organizations, as both service clubs and labour unions are there to protect the interests of the members, sometimes even at the expense of other members of society. Since both of these organizations require a heavy level of bargaining at regular intervals, the members do need to be aware of the overall circumstances, gains and losses by similar organizations, and the options available to them. As a result, my suspicion is that the nature of the news and affairs attended to by members of such organizations will also have a self-interested nature, and are not necessarily the news and affairs of interest by the general public.

Conclusion

To recap what we have learned so far: 15 different components have surfaced, each looking at a particular facet of social capital. This shows a more elaborate picture, compared to what was suggested by Putnam, but it does not yet answer the key question of whether the resultant dimensions are all parts of a one bigger package. That will be explained in the next two chapters. To make the discussions in those chapters easier to follow, it helps to name the 15 dimensions, keeping in mind that the proposed names are only to facilitate the later recollection of various components, and not to comply with any pre-defined standard. Here is the list of the proposed names: 1) *Trust*; 2) *Confidence in public institutions*;

3) *Voting*; 4) *Religious Engagement*; 5) *Volunteering*; 6) *Political Party Involvement*; 7) *Neighbourliness*; 8) *Political sensitivity*; 9) *Confidence in private-sector institutions*; 10) *Engagement in recreational activities*; 11) *Irregular political activism*; 12) *Informal social networks*; 13) *Cultural-communal engagement*; 14) *Donation*; and 15) *Social engagement for self-interest purposes.*

5 Social Capital and Regional Diversity: The Provinces

Discussions of national and international trends of social capital, like those offered in the first couple of chapters of the book, may leave the impression that social capital is a national entity, and that each country is homogeneous in terms of its social capital character. In reality, however, the level and nature of social capital within each country varies greatly depending on region, city, neighbourhood, ethnic group, age group, and so on. Two studies by Putnam (1993; 2001) illustrated this point clearly; the former showed the distinct social capital profiles of northern versus southern Italy, and the latter the state variations in the distribution of American social capital. An understanding of social capital trends in a country, therefore, would be incomplete without a grasp of such internal variation.

One good reason for studying the regional distribution of social capital has to do with the policy-making potential of such a study. While the national profiles are normally shaped through historical processes that are of *longue durée* and, hence, less subject to short-term policy-making efforts, the regional dynamics are influenced by such policies to a greater extent. In other words, the regional picture would provide the policy-makers with targets toward which they can direct their efforts. This is more so in countries in which there are strong provincial jurisdictions.

The regional and provincial differences in Canada are known facts. The later development of western Canada, for instance, put a few of the western provinces on a different development path compared to the eastern parts of the country. Besides such historically different experiences, the waves of economic recession that have hit the country since the early 1980s have affected various regions differently. Finally, the long-standing issue of the separatist sentiments in the province of Quebec has created a totally different political dynamic and social environment in that province compared to the rest of Canada. All of such developments could have potentially important consequences for the state of social capital in different regions of Canada.

As a step in that direction, the present chapter looks into the social capital profiles of various provinces within Canada. The main purpose here is to arrive at a big picture, rather than to delve into the specific circumstances of each province and the connections between those circumstances and their social capital attributes. Towards this goal, I have examined the provincial distribution of the 15 social capital dimensions identified in chapter four. As the readers would certainly remember, some of those dimensions had more clear-cut boundaries and conceptual clarity than others; and, by virtue of their higher clarity, those dimensions will act as more useful and informative conceptual tools for seeing provincial differences, and should accordingly be given more thought in future research.

Provinces and the Dimensions

Let's start with examining the least problematic dimensions, that is, those for which there appeared no significant variation among provinces. These dimensions consist of political party involvement, irregular political activism, political sensitivity, and engagement in self-interest activities. The provincial distribution of the values for these components of social capital, reported in Figures 5.1 through 5.4, indicates that most of those values fluctuate around the national average. Of these, the values reported for political sensitivity and engagement in self-interest activity are virtually identical, but those of irregular political activism and political party involvement show a bit of variation, with slightly higher values reported for B.C. and the Atlantic provinces for the former, and Saskatchewan and the Atlantic provinces for the latter.

Figure 5.1: Political Party Involvement, 2003

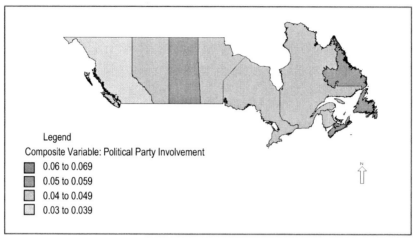

Source: GSS17 Masterfile

Figure 5.2: Irregular Political Activism, 2003

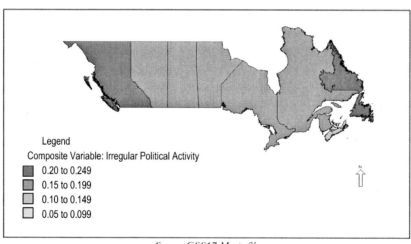

Source: GSS17 Masterfile

Figure 5.3: Political Sensivity, 2003

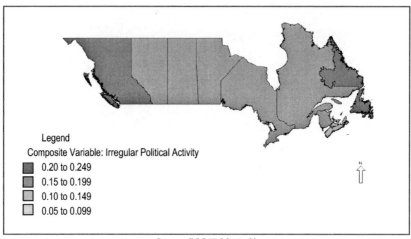

Source: GSS17 Masterfile

Figure 5.4: Engagement in Self-Interest Activities, 2003

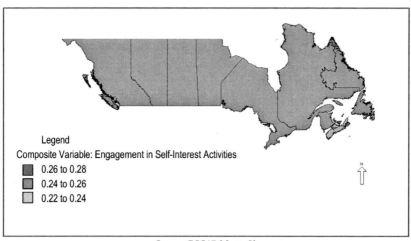

Source: GSS17 Masterfile

Figures 5.5 through 5.12 show the provincial distribution of a different set of social capital dimensions: trust, neighbourliness, social networks, cultural-community engagement, religious engagement, donation, volunteering, and engagement in recreational activities. The common feature of the provincial distribution of these graphs is that the values reported for Quebec and Ontario are noticeably lower, and those reported for the rest of Canada are higher, than the national averages. Also, in most cases, Quebec is behind all other provinces by a wide margin. This pattern is to some extent similar to what Putnam (2001) found for different states in the U.S. – i.e., social capital index rises as one moves from east to west – except that in the case of Canada the indexes rise as one moves from centre towards both west and east coasts.

Figure 5.5: Trust Composite Index, 2003

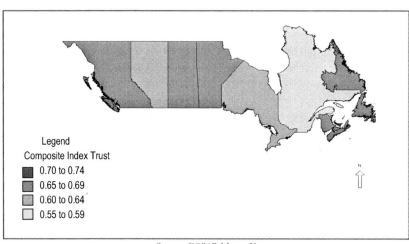

Source: GSS17 Masterfile

83

Figure 5.6: Neighbourliness, 2003

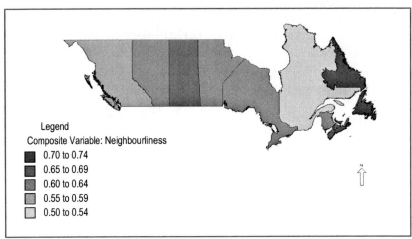

Source: GSS17 Masterfile

Figure 5.7: Social Networks, 2003

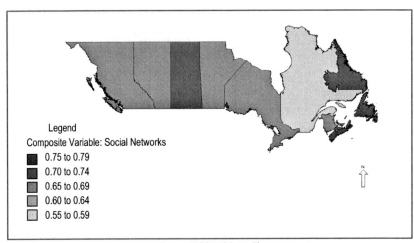

Source: GSS17 Masterfile

Figure 5.8: Cultural Community Engagement, 2003

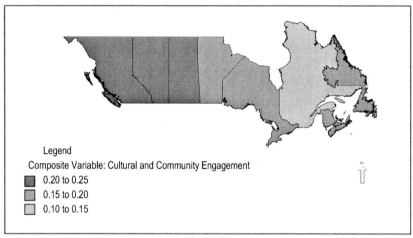

Source: GSS17 Masterfile

Figure 5.9: Religious Engagement, 2003

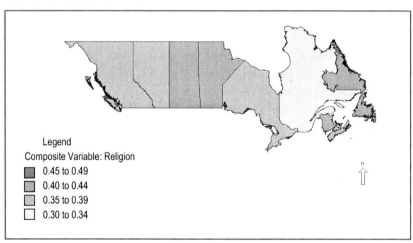

Source: GSS17 Masterfile

Figure 5.10: Donation, 2003

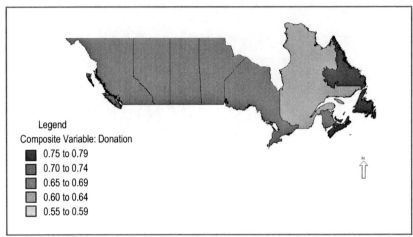

Source: GSS17 Masterfile

Figure 5.11: Volunteering, 2003

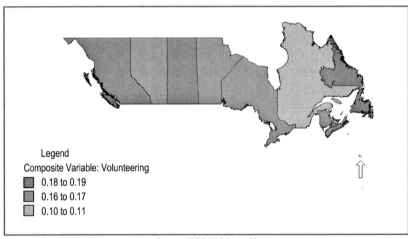

Source: GSS17 Masterfile

Figure 5.12: Engagement with Recreational Activity, 2003

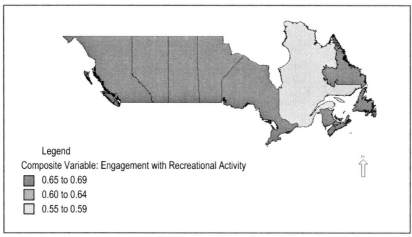

Source: GSS17 Masterfile

There is yet a third group of social capital dimensions for which the dominant pattern is the exact opposite to what was seen for the previous set, in that the value reported for Quebec is noticeably higher than that of any other province. This is true for confidence in public institutions, voting, and confidence in private corporations, as illustrated by Figures 5.13 through 5.15. For the latter two dimensions, the Atlantic provinces also demonstrate a higher value compared to the western provinces.

The information presented above can be used to conceptualize many legitimate and exciting questions for further research, addressing all of which is obviously beyond the present study that is concerned with arriving at the general picture. What we can do here, however, is to speculate about some possible factors behind the observed trends, as guide for future research.

In a general way, the 15 dimensions of social capital can be clustered into three broader groups: 1) those dealing with the elements and building blocks of community; 2) those dealing with government and public institutions; and, 3) those revolving around the private sector. Re-examining the above maps along these conceptual lines points to a unique status for Quebec, in which a weaker set of communal ties is combined with a higher degree of confidence in both public and private sectors.

87

Figure 5.13: Confidence in Public Institutions, 2003

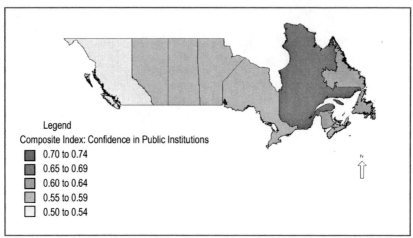

Source: GSS17 Masterfile

Figure 5.14: Voting, 2003

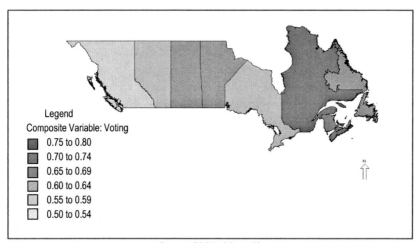

Source: GSS17 Masterfile

Figure 5.15: Confidence in Private Institutions, 2003

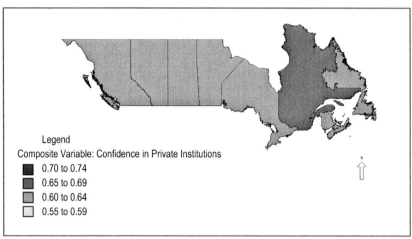

Source: GSS17 Masterfile

Can this be related to the presence of strong sovereignty sentiments in the province? Given that, in their struggle for getting an acknowledgement of their distinct identity or independence, Quebecers have mostly acted through voting and have had the support of the provincial government on their side, is it reasonable to assume that this has raised their level of confidence in governmental institutions? As for the higher level of confidence they have expressed in the private sector institutions, can we attribute that to the presence of a stronger individualist outlook, an outlook also strongly present among the French in France? How about the relatively weak sense of belonging to a community in Quebec, reflected in a noticeably lower level of social trust among the province's residents, a smaller size of their social networks, a weaker sense of neighbourliness, smaller amount and frequency of donation, and a poor tendency for volunteer activities?

One possible explanation for the latter phenomenon – weaker communal ties – can be the dual language composition of the province. Existing research has shown the significance of a common language in facilitating the expansion of social networks, participation in community projects, volunteering, and, finally, promoting interaction and long-lasting social connections with others; these, in turn, would result in a more trustful relationship. The segmentation of a community along language lines would limit each group to opportunities available only in their own segment, hence, an overall lower score for the whole community.

A second possible explanation of the unique status of Quebec would revolve around the impact of the presence of pro-sovereignty sentiments among the Francophone residents of the province, and lack thereof among the non-Francophone ethnic groups. Given that most of latter groups seem to be against Quebec's separation, their votes in the referendum and other elections could have been seen by the Francophone majority as a hurdle in way of the success of the sovereignty movement. The presence of such sentiments was well captured by the now-famous statement by Jacques Parizeau, then the Premier of Quebec, after the defeat of the sovereigntists in the last referendum in 1995 when he said, "It's true, it's true that we have been defeated, but basically by what? By money and some ethnic votes, essentially" (CBC, 2006).

A third possible explanation for Quebec's unique social capital status – and its low level of trust, in particular – may be the presence of a low-trust culture among the French. The essence of this hypothesis is that there exists a correspondence between the social capital tendencies of each ethnic and cultural group in Canada and that of the society they have come from or the culture they belong to. In the case of Quebec, this linkage can be viewed to exist between the social capital attributes of the French-speaking population in Canada and that of the French in France.

Conclusion

The examination of regional distribution of various social capital indicators revealed a distinct status for the province of Quebec, in which a stronger confidence in public and private sectors was combined with weaker communal ties. Three possible explanations for this unique status were offered. However, an empirical examination of those explanations requires the availability of data sources other than the ones currently in existence. The existing data allow us to partially address some of the questions raised above. Before doing that, however, the next chapter provides a comparison of immigrants and native-born Canadians along the fifteen dimensions of social capital.

6 Social Capital and Immigration Status: Immigrants and the Native-born

There are many different faces of diversity in any given country. As far as ethnic and cultural diversity is concerned, we can think of two broad types: the old faces of diversity, or those reflecting the permanent and historically-shaped composition of the populations; and, the new faces, resulting mostly from immigration. In Canada, both types of diversity are present; the former type is represented by the presence of Francophone and Aboriginal populations; the latter is caused by immigration. A study of the interaction of diversity and social capital, therefore, needs to address both of these aspects.

Our discussion in chapter five revolved around the regional distribution of social capital in Canada. That discussion showed a distinct status for Quebec compared to the rest of Canadian provinces. Given the concentration of the Francophone population in Quebec, such provincial profiles can act, to some extent, as proxies for older faces of diversity. In this chapter, I add the new source of diversity – that is, immigration – to the picture, by comparing the immigrant and native-born sub-populations along the fifteen dimensions of social capital identified in chapter four.

There are some good reasons to expect the social capital profile of immigrants to differ from that of the native-born Canadians. First, most recent immigrants to Canada are now coming from less-developed societies and carry with them residuals of their home cultures, and in particular, their emphases on communal ties. Given the drastic differences in the states of social capital in different countries, one would expect that such differences get carried into host societies through migration. Second, on top of the initial differences in the social capital tendencies of immigrants and the native-born, there exists the possibility that immigrants' socioeconomic experiences in host societies – which are distinct from those of the native-born – would also trigger different social capital tendencies among them. Some immigration researchers, for instance, have speculated that many immigrants experience a sudden drop in their personal resources upon arrival and find no other recourse but to rely on their communal resources for survival (Portes, 1995). Viewed against the backdrop of a prevalent individualistic orientation in many immigrant-receiving countries, this might suggest that the stock of social capital is disproportionately stronger among minorities.

For immigrant-receiving countries, a thorough social integration of immigrants into mainstream society is a big concern, and many aspects of such an integration involves variables and processes we have been discussing in this book under the rubric of social capital. For example, whether or not immigrants have confidence in public institutions is related to whether or not such institutions operate in an unbiased and accommodating fashion; whether or not they trust others in society is influenced by the extent to which they have been treated respectfully and without prejudice; whether or not they enter into social networks with people other than their co-ethnics depends of how welcoming those others have been; and so on.

Immigrants and the Native-born: The Social Capital Profiles

Let us start with the dimensions for which immigrants and the native-born are not drastically different; this includes trust, political sensitivity, confidence in private institutions, political party involvement, cultural-communal engagement, and engagement in self-interest activities. Figure 6.1 shows the proportion of immigrants and non-immigrants living in major Canadian cities who believe that others can be trusted. The values indicate that, except for one or two cities, there is no noticeable difference between the two groups. Later in this book, we will compare the trust levels of each immigrant group with those of their countries of origin, but it seems that even if they inherit a lower level of trust from their initial home, this gets balanced out with the positive impacts of education and/or other demographic factors. The second and third dimensions in this league are political sensitivity (Figure 6.2) – which measures the tendency to acquire information on political issues, contacting newspapers to express an opinion, speaking at a public meeting, and helping others through teaching, coaching or practical advice – and confidence in private-sector institutions (Figure 6.3). Figures 6.4 through 6.6 show relatively similar, but low, levels of engagement in political, cultural, and trade-union-related activities for both immigrants and the native-born.

Figure 6.1: Trust, by immigrant status and city

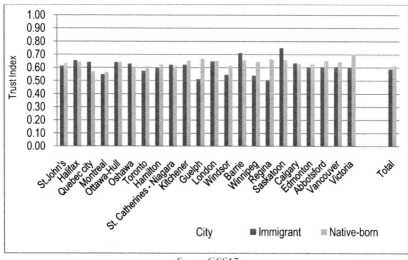

Source: GSS17

Figure 6.2: Political sensitivity, by immigrant status and city

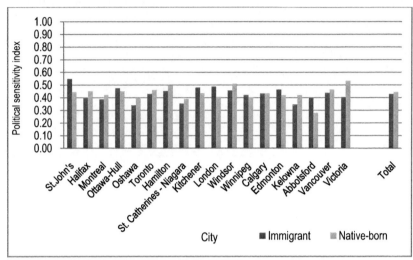

Source: GSS17

Figure 6.3: Confidence in private institutions, by immigrant status and city

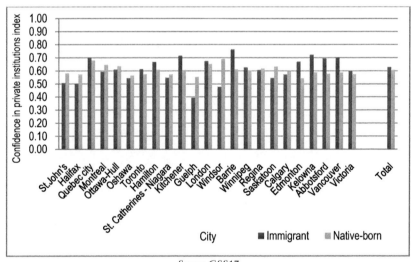

Source: GSS17

Figure 6.4: Political party involvement, by immigrant status and city

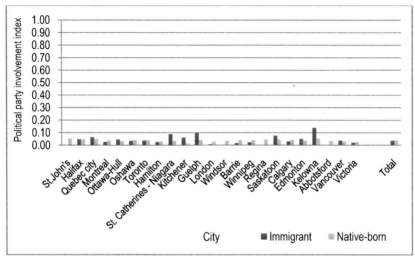

Source: GSS17

Figure 6.5: Cultural communal engagement, by immigrant status and city

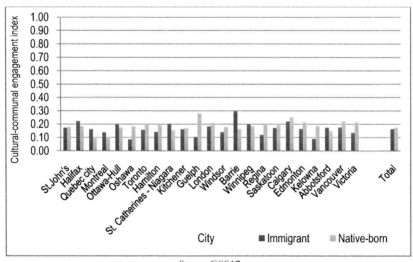

Source: GSS17

Figure 6.6: Engagement in self-interest activities

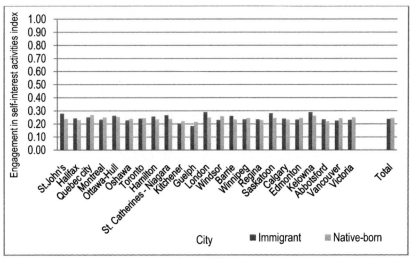

Source: GSS17

The second group of variables consists of two dimensions in which immigrants have a higher than average value than the native-born; this includes confidence in public institutions (Figure 6.7) and religious involvement (Figure 6.8). The reasons for immigrants' higher scores in these two areas are easy to grasp. In terms of religious engagement, many recent immigrants come from countries that are less secular than Canada; for them, religion has a much more visible role in their daily functioning. Moreover, after arrival, many such immigrants find themselves in a situation of lost or shrunken social networks, in which religious communities can provide a counter-balance. In this kind of circumstance, even the secular immigrants might seek refuge in the collectivity that religious bonds can offer to them.

Figure 6.7: Confidence in public institutions, by immigrant status and city

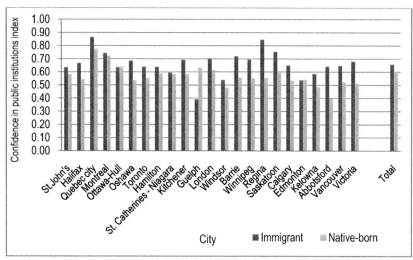

Source: GSS17

Figure 6.8: Religious engagement, by immigrant status and city

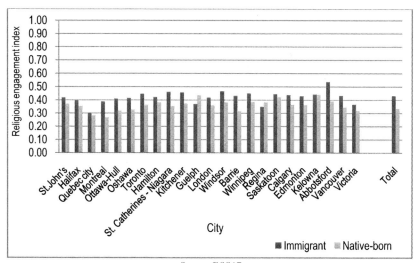

Source: GSS17

As for immigrants' higher confidence in public institutions – such as health care, education, judiciary system, and so on – a contrast with the status and effectiveness of such institutions in their home countries could be the cause. In many developing countries, the public institutions do not function as effectively and rationally as they do here in Canada. If nothing else, the functioning of such institutions in Canada demonstrates a higher degree of stability, more accountability, and less corruption than their home countries, all good reasons for immigrants to attribute a higher score to them.

A series of interviews conducted with immigrants from a developing country immediately before their migration revealed that the perceived better performance of public institutions in the areas of education, bureaucracy, health care, and security is a main reason for their decision to migrate. One such immigrant, for instance, had the following to say:

> The biggest advantage that Canada has for me is its rule of law. Based on what I have heard, in Canada, if you want to do something, it is either possible or not, and when it is, it is so for everyone.... . Here, everything is constantly changing.... . Law changes on a second-by-second basis.... . (quoted in Kazemipur, 2004: 34)

Another interviewee, who had spent some time in the U.S. before, and was now planning to migrate to Canada, pointed to the functioning of public institutions more clearly:

> I don't know and I don't care what Clinton does. He may order a thousand people to be killed in the other end of the world, but what you and I see on a daily basis, and what is so obvious to us, is that when you go to, say, municipality office [City Hall], your job is taken care of easily… or you can take care of a lot of things only through phone… when I went to the U.S. for the first time, what surprised me a lot was that nobody in any of those offices would say 'no' to you…they even offered other and better ways to fix a problem – This is what matters: the way you are treated on a daily basis; this is what is so bad here [in my home country]– hospitals, municipalities, courts, law enforcement agents, etc. What politicians do up here does not hurt you directly, but the police agent who is supposed to be there, say, after you had a car accident, and he does not show up for three hours – that is what hurts you. (quoted in Kazemipur, 2004: 39)

Religious engagement and confidence in public institutions aside, there are several dimensions of social capital in which the native-born show a higher average value (Figures 6.6 through 6.15). Some of these dimensions – such as voting and engagement in irregular political activities – require citizenship status; others – like social networks, neighbourliness, and volunteering – are contingent upon a long-time presence in the community, the formation of adequate social ties with the mainstream institutions, and the development of adequate language skills. Yet, a third set – donation, and engagement in recreational and communal activities – requires a certain degree of economic and financial stability.

Figure 6.9: Voting, by immigrant status and city

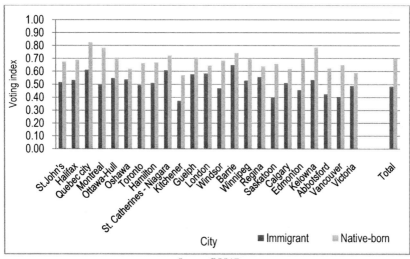

Source: GSS17

Figure 6.10: Volunteering, by immigrant status and city

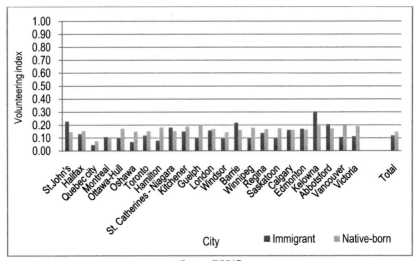

Source: GSS17

Figure 6.11: Neighbourliness, by immigrant status and city

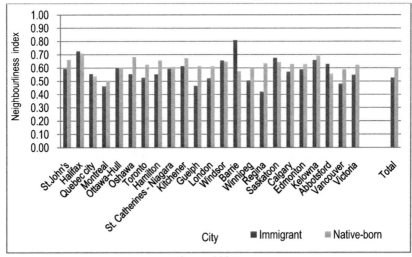

Source: GSS17

Figure 6.12: Irregular political activism, by immigrant status and city

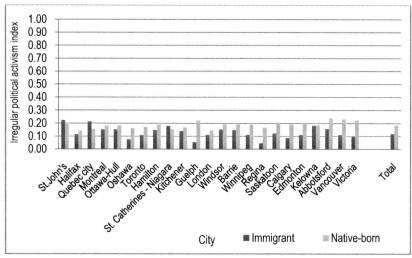

Source: GSS17

Figure 6.13: Social network, by immigrant status and city

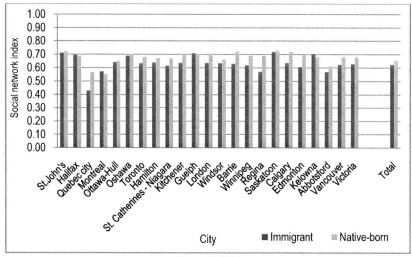

Source: GSS17

Figure 6.14: Engagement in recreational activities, by immigrant status and city

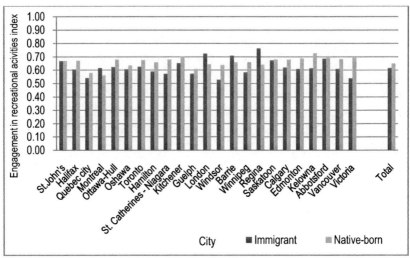

Source: GSS17

Figure 6.15: Donation, by immigrant status and city

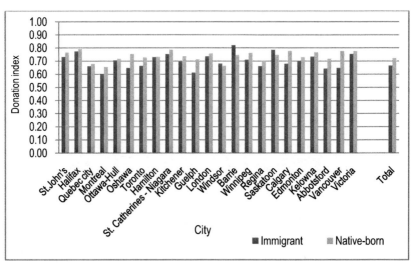

Source: GSS17

Conclusion

The big picture conveyed by the above figures makes sense, for the most part. The ambiguities that might exist with regards to some of these trends have mostly to do with the ambiguities in the nature of those dimensions. As mentioned in chapter four, some of the resultant dimensions of social capital have a less-than-desired level of conceptual clarity, which poses challenges for the interpretation of how these dimensions behave. Notwithstanding this shortcoming, the trends observed above provide a rich set of information for future research to focus on.

One area that can reveal a great deal about the above dynamics is inter-city variation. As the reader has undoubtedly noticed, despite the average index scores for immigrants and non-immigrants, the patterns are not consistent for all cities compared. Those cities that show unexpected patterns are perfect sites for future research in order to further our understanding of how social capital dimensions interact with other contextual factors present in those cities. In the next section of the book we take on this task through a heavier focus on trust.

PART III:

Diversity and Trust

7 Trust and Diversity in Cities

The discussion of the various dimensions of social capital in the previous section showed the salience of trust as one of the major components. Indeed, the significance of trust in facilitating the functioning of other dimensions of social capital has been so great that the initial studies of this subject used trust as the surrogate for social capital. Without trust, Warren (1999: 2) argues, "... the most basic activities of everyday life would become impossible." Along similar lines, Putnam (2000: 135) points out that, "[h]onesty and trust lubricate the inevitable frictions of social life." Due to the centrality of the concept of trust, the focus of the remainder of this book will be on the interaction of trust and diversity.

Why Trust?

Despite some other initial works on trust (see, for instance, Coleman, 1990; Gambetta, 1988; 1993), Francis Fukuyama was the first scholar who gave visibility to the concept, particularly in relation to the economy (Fukuyama, 1995a; 1995b). In his *Trust: The Social Virtues and the Creation of Prosperity*, Fukuyama (1995b) bases his argument on two premises; first, that the significance of culture in global affairs has been rising in the post-Cold War era, an argument made also by some other scholars like Huntington (1996); and, second, that one area in which this rising influence is most visible is the realm of economic development. According to him, one major predictor of a country's economic performance is the degree to which its national culture is conducive to economic prosperity or acts as a facilitator for it. In this perspective, the presence of high trust, as one of the key components of culture, would have an extremely positive impact on the functioning of the economy, in that it leads to a lesser need for what economists calls 'transaction costs' – i.e., the costs of setting up elaborate formal and legal procedures to monitor the economic transactions among firms and individuals. A lower transaction cost, in turn, reduces the expenses associated with economic activities and results in a more vibrant and flexible economic machinery.

Against this conceptual background, Fukuyama (1995b) introduced a continuum on which countries are located according to their levels of what he called a 'generalized social trust'. He then argued that the location of countries on this continuum directly corresponds with their potential for economic prosperity. At one end of this spectrum was a country like Russia, having the lowest level of trust, and at the other end, countries such as Germany and Japan, with highest levels of trust and economic dynamism. Next to Russia were Taiwan, Hong Kong, and China, followed by France and Italy; and next to Japan and Germany were South Korea and the United States. Fukuyama then provided a very elaborate account of the mechanisms through which trust, or lack thereof, influences not only the overall performance but the also the nature of the economy in each country.

The impact of trust on politics has also received some attention. In his *Making Democracy Work*, Putnam (1993) drew attention to the positive correlation between trust and the democratic nature of local governance structure in various parts of Italy. In another study, Inglehart (1999) showed, drawing on the data from the World Values Surveys across 41 countries, that the long-term stability of existing democracies is heavily influenced by subjective well-being and interpersonal trust.

One of the useful contributions of these studies is the distinction made between different types of trust. Warren (1999), for instance, distinguishes between *social trust* – among citizens – and *political trust* – between citizens and government. Those concerned with social trust, in turn, have made a distinction between *particularized* and *generalized* trust – trust in people we know well versus trust in general and anonymous others, respectively. The subjects of particularized trust are typically family members and close friends, with whom one normally has long-lasting and multifaceted relationships. The duration and multiplicity of such relationships make them favourable grounds for development of trust; but, paradoxically, this is exactly what makes particularized trust less important. There is almost a consensus among researchers that it is generalized trust – that is, trust in strangers – that is most conducive to both a healthy economy and a vibrant democracy (see, for instance, Fukuyama, 1995a; 1995b; Inglehart, 1999; Offe, 1999; Putnam, 2000; Uslaner, 1999).

This latter type, generalized trust, is the focus of the remainder of the book, starting with an examination of the interaction of trust and ethnic/cultural diversity. Studying trust dynamics in smaller geographic units such as the city, while an enlightening exercise, comes with its own caveats. For one thing, it requires a conceptual perspective different from one used to analyze trust among individuals, which is what our minds are immediately used to. This distinction is important to be aware of, because the factors at work at the individual level are not necessarily useful or even relevant for the analysis of larger entities such as cities. For example, speaking of the relationship between income inequality and trust is meaningful for a city, but not relevant if the focus is on why a particular person trusts or distrusts.

A second caveat in using smaller geographical units for the analysis is that it inevitably results in the study of smaller populations, which could become problematic if their sizes fall below a reasonable threshold. Given that most survey data sources are based on samples rather than entire populations, this can make the generalization of the findings more legitimate for some groups than others. No matter how comprehensive the sampling designs of such surveys could be, when it comes to very small units, low counts and reduced reliability of the results seem to be inevitable consequences. Due to this limitation, some of the Canadian cities with noticeably small number of cases in the samples used have been dropped from the analysis.

This chapter has three main parts. First, a quick snapshot of the distribution of trust in Canadian cities will be provided. Second, the main factors influencing the trust level of cities will be briefly discussed. Third, the discussion of those factors will be used as a conceptual framework to develop hypotheses about the level of trust in Canadian cities, the validity of which will be examined using the Canadian General Social Survey cycle 17 data.

The City Differences: How Much and Why?

Figure 7.1 shows the proportion of each city's population who believe that people can be generally trusted. The overall profile of the graph indicates that the residents of the Canadian cities are, by and large, trustful; many cities reach high values in the range of 60 and 70 percent. At the same time, there are a few cities with their trust levels below 40 percent, a value almost half of those reported for the high-trust cities. While most of the low-trust cities fall in the province of Quebec – as expected from the discussion in the previous chapter – there are some fluctuations in trust levels of cities outside Quebec as well that warrant an explanation.

Figure 7.1: Trust Level by City, 2003

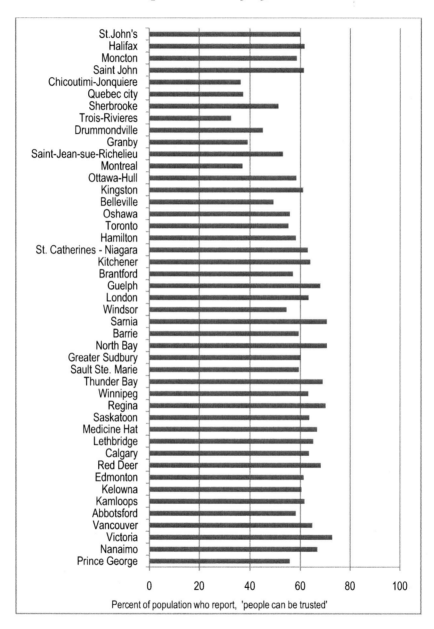

Addressing the factors influencing a city's trust level, and relying on the literature on trust, Figure 7.2 highlights five possible variables: population of the city, population of immigrants as a percentage of the total city population, the average income of city dwellers, the degree of deprivation and income inequality in each city and, finally, the extent to which the city is ethnically diverse. Below, we explain the nature of the possible relationships between these variables and trust.

Figure 7.2: Determinants of Trust at City Level

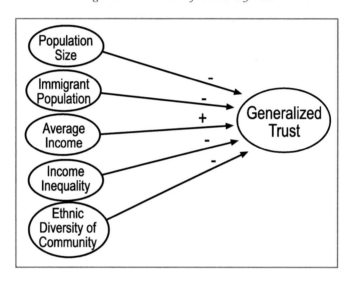

a. Population size

Putnam (2000), among others, has shown that with an increase of population, the level of trust normally drops. This has to do with the fact that in almost every society, large cities create a structural setting that is radically different from that of small towns, in that the former possess more diverse occupational structures, higher population density, faster life tempo, and more anonymous crowds. As far as trust is concerned, with the increase of population, social relationships and interactions among city dwellers tend to become shallower and shorter, robbing them of a deep and reliable knowledge of each other. This, in turn, raises the risks associated with trusting others, leading to a weaker propensity to trust.

b. Average income

Wuthnow (2002) has argued that one of the reasons behind the declining levels of trust in American society is the decline of the economic resources of Americans. The logic of this argument is not too difficult to grasp. Trust involves risk, and those who are in better positions to survive those risks are more likely to trust. In contrast, those with limited resources may withdraw from trusting others, simply because the materialization of any such risks could not be easily compensated for (Hardin, 1999).

c. Income inequality

Inequality can affect trust levels in two ways: it can diminish the resources available to lower classes, hence, lowering their tolerance for the risk associated with trust; and, it can generate a sense of alienation and indifference towards others, which can lead to an elevated sense of distrust. In general, therefore, the uneven distribution of material resources can result in a lower level of trust. Here, we have used the proportion of the city's population who live in poverty as an indicator of the extent of deprivation and inequality in the city.

d. Immigrant population

Some previous studies have indicated that immigrants typically express a lower level of trust compared to the native-born populations (Soroka, Helliwell, and Johnston, 2003; Rice and Feldman, 1997). One possible reason for this may be their minority status, which can block or slow down their integration into the mainstream population. Such exclusion can then result in lower levels of contact between them and the majority, resulting in the presence of no or a very limited history of past interactions between the two groups. The paucity of past interactions, in turn, reduces the ability of each group to accurately predict the reactions of the other party, thus, raising the risks associated with trusting anonymous others. If this hypothesis happens to be valid, it would basically mean that with an increase in the immigrant population, a city's overall level of trust should drop.

e. Ethnic diversity

The existing research on the possible relationship between diversity and trust is inconclusive, but the overwhelming majority of the studies done so far argue in favour of a negative relationship between the two, that is, the more diverse a population, the lower its overall trust level (see, for instance, Putnam, 2003; Alesina and La Ferrara, 2002; Knack, 2001; Knack and Keefer, 1997; Zak and Knack, 2001; Wuthnow, 1998). The reason for the lower level of trust in more ethnically diverse environments can be related, as some have pointed out, to the lower level of trust among those who put their ethnic identities first. In a comparative study of Canada and the U.S., for instance, Helliwell (1996) finds that those Canadians who qualified their citizenship with French, English, or Ethnic had lower levels of trust than those who considered themselves Canadian first or only; and that a similar trend exists for Americans as well. As a result, the overall score for the whole city can be dampened by the presence of more low trusting individuals.

But, there is more to the story than this simple arithmetic. The increase in the diversity of an environment can also trigger a change in the attitudes of initial residents, not only towards people who are different but also towards people of similar background. Putnam (2007), for instance, has shown that in racially diverse neighbourhoods residents of all races tend to 'hunker down'; "trust (even of one's own race) is lower, altruism and community cooperation rarer, friends fewer (p. 137)." The opposite scenario has also been shown to be at work. In a study of the Detroit dwellers, for instance, Marschall and Stolle (2004) found that "... neighbourhood racial heterogeneity ... significantly increase[s] blacks' propensity to trust others" (p. 146). What is more interesting about their findings, however, is that the racial heterogeneity of a neighbourhood does not work the same for Blacks and Whites. The latter finding alludes to the possibility that diversity and trust may interact differently in different countries and for different populations.

The Results

In the next few pages, the validity of the above hypotheses in a Canadian context will be examined (the data sources for all variables involved are reported in Appendix 7.1, and the correlation coefficients for this and all other variables in Appendix 7.2). Figure 7.3 indicates that, as hypothesized, there is a negative correlation between population size and trust level, that is, the cities that are less crowded tend to have higher levels of trust; and as the population increases, the trust level generally drops. The negative nature of the relationship between the two variables in also reflected in negative value of the correlation coefficient (-0.19). It should be noted, however, that the reported value is not statistically significant, partly due to the weakness of correlation and partly due to the limited number of cities included.

Figure 7.3: Trust Level by Population, Canadian Cities, 2003

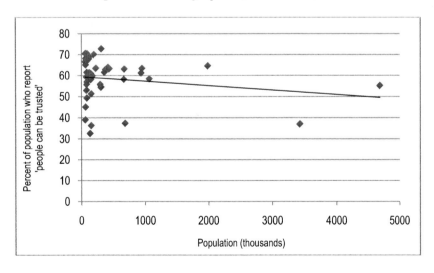

Figure 7.4 shows the relationship between income and trust levels in Canadian cities. As expected, the trust level rises as the average income in the city increases. The correlation coefficient reported for income (0.32) is positive, moderately strong, and statistically significant. Although the relatively small number of cities in the analysis does not allow for much empirically-based theorizing, it would be useful to point to a potentially interesting feature in the distribution of cities. This feature, which is represented by the dotted line in Figure 7.4, points to the possibility of a curvilinear relationship between income and trust, in that the trust level of city increases only as their average income rises from a low to a middle level; after that, it either remains the same, or declines. This can suggest that the psyche and cultural outlook of those living in very prosperous and fast-growing cities might become less trust-inclined, possibly due to the fast pace of life and the predominance of a materialistic outlook. In other words, a healthy state of social trust seems to be associated with a middle average income, rather than extremely low or high incomes.

Figure 7.4: Trust Level by Average Income, Canadian Cities (only CMAs)

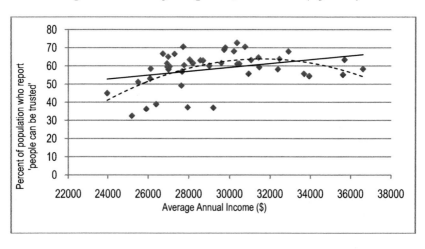

Figures 7.5 through 7.7 illustrate the relationship between trust and economic deprivation, as measured by the proportion of a city's population

who fall under the low-income cut-off line. As hypothesized, the relationship between the two variables is negative, meaning that with higher levels of poverty in a city, there will be a lower level of trust in others. This negative correlation between poverty and trust is consistent for all three types of poverty measures – families, households, and individuals – producing correlation coefficients of -0.37, -0.36, and -0.49, respectively, all statistically significant. Given the relatively small number of cities involved in the analysis, the statistical significance of the resultant coefficients testifies to the strength of the relationships.

Figure 7.5: Trust and poverty (of individuals) in Canadian Cities

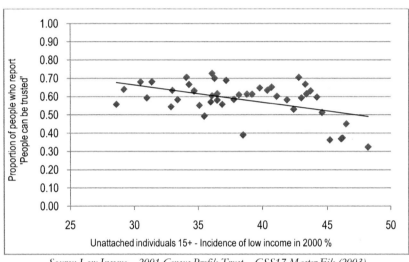

Source: Low Income – 2001 Census Profile Trust – GSS17 Master File (2003)

Figure 7.6: Trust and poverty (of economic families) in Canadian cities

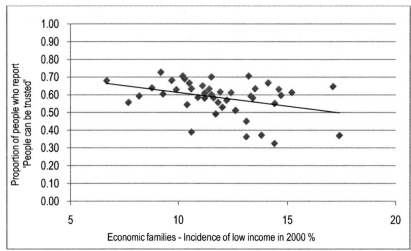

Source: Low Income – 2001 Census Profile Trust – GSS17 Master File

Figure 7.7: Trust and poverty (of households) in Canadian cities

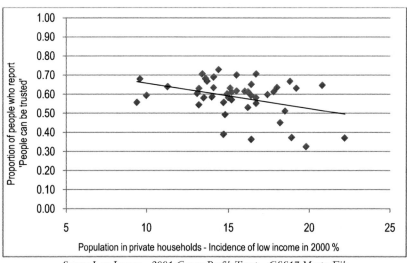

Source: Low Income – 2001 Census Profile Trust – GSS17 Master File

Figure 7.8 illustrates the relationship between trust level in a city and its degree of ethnic/cultural diversity, measured by the Index of Qualitative Variation (IQV). The technical description of this measure is included in Appendix 7.3; here, it suffices to mention that IQV can take any value between 0 and 1, with the former indicating the presence of no diversity, and the latter that of perfect diversity. The graph indicates that cities with more ethnically diverse populations tend also to demonstrate a higher level of trust. This runs contrary to the bulk of research on the relationship between diversity and trust cited earlier, particularly the study by Putnam (2003), in which he found a negative relationship between the two variables in American states. The correlation coefficient reported for this pair of variables is 0.75, a strong positive association, which is also statistically significant.

Figure 7.8: Trust Level by ethnic diversity, in Canadian Cities

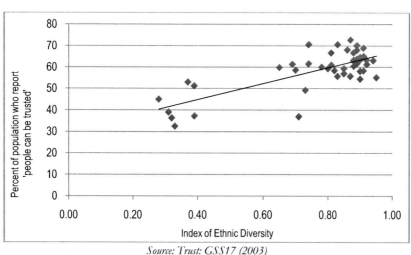

Source: Trust: GSS17 (2003)
IQV: Author's calculations based on census 2001 data

The unexpected nature of the relationship between ethnic diversity and trust level in Canadian cities warrants a closer examination in order to establish that it is an indication of a genuine relationship, and not a statistical artefact. One possibility is that the above-mentioned trend has been heavily influenced by cities in Quebec, in which both the ethnic diversity and trust levels are extremely low, noticeably lower than those for all other cities (see Figures 7.9 and 7.10).

Figure 7.9: Trust by city, 2003

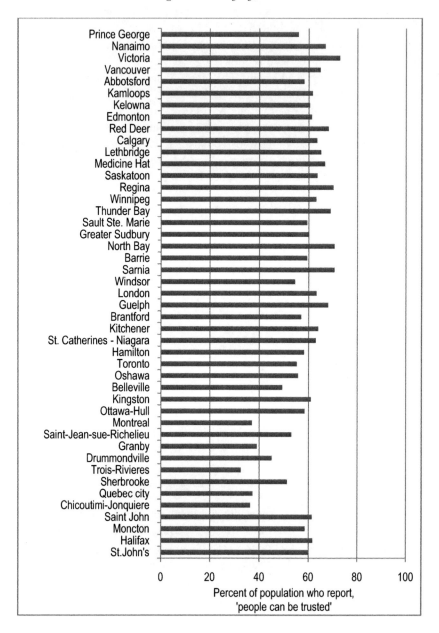

Figure 7.10: Ethnic/cultural diversity of Canadian cities, 2001

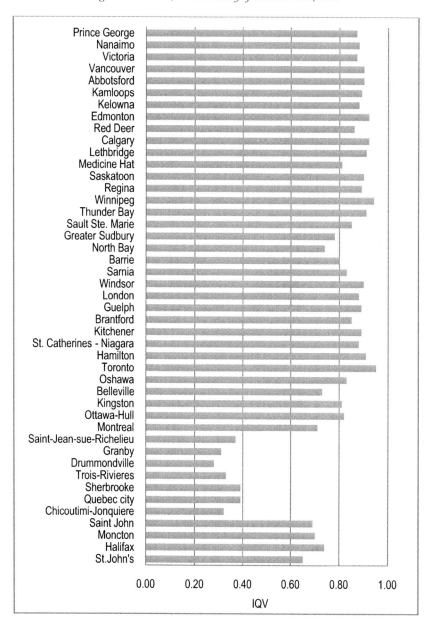

In other words, it is theoretically possible that the unexpected relationship between these two variables is due to a combination of low-trust and low-diversity in Quebec, and high-trust and high-diversity in the rest of Canada. Distinguishing these two groups of cities will allow us to examine the validity of this theoretical possibility.

Figures 7.11 and 7.12 show the relationship between trust and diversity, separating cities in Quebec from those in the rest of Canada. The separation of these two clusters results in two distinct patterns, a negative correlation for cities in Quebec (-0.13) and a positive but weaker relationship (0.17) for the rest of the Canadian cities. It should be noted also that both of these coefficients are now statistically non-significant, a reflection of a weaker association and a smaller number of cities in each group. Although this implies a difference between Quebec and the rest of Canada, it does not suggest that the unexpected positive correlation between diversity and trust in Canadian cities is influenced by a 'Quebec exceptionalism.' If anything, the negative correlation of these two variables in Quebec is in tandem with what the existing literature suggests, and it is the rest of Canada that constitutes an exception in this regard. Let's dig a little deeper into this before jumping to any final conclusion.

Figure 7.11: Trust level by ethnic/cultural diversity, Canadian Cities excluding Quebec

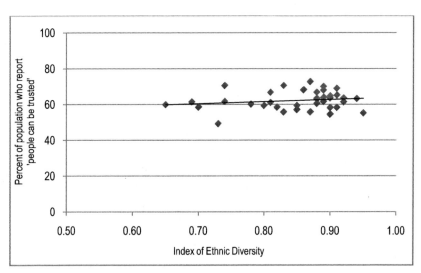

Figure 7.12: Trust level and ethnic/cultural diversity, Quebec Cities

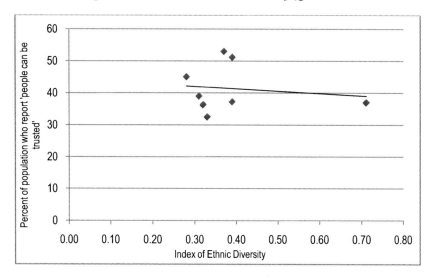

Figure 7.13: Trust level by ethnic/cultural diversity, Quebec Cities excluding Montreal

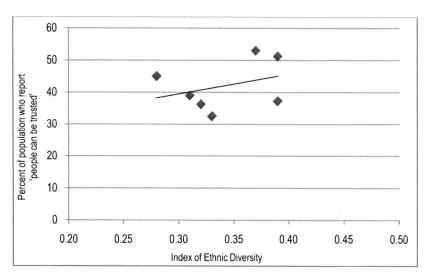

125

A closer examination of the graph in Figure 7.12, however, shows that the overall trend of diversity negatively correlating with trust has been heavily influenced by one city, Montreal, located at the right half of the graph. The noticeable distance between the rest of cities in Quebec and Montreal raises the possibility that the latter acts as an outlier as far as the relationship between trust and diversity is concerned. The picture might be further refined if this anomaly is taken out and the patterns are revisited. Figure 7.13 shows the revised pattern for Quebec after excluding Montreal. As the trend line shows, as a result of excluding Montreal, the nature of the relationship between the two variables in Quebec cities changes drastically, from negative to positive (from a value of -0.17 to 0.34), making it consistent with the trend observed for the rest of the Canadian cities.

Given this, it would make sense to run the analysis once more for all Canadian cities, including cities in Quebec but excluding Montreal, and re-examine the nature of the relationship between diversity and trust. Table 7.1 provides the summary of results for all the different scenarios discussed above. It shows that excluding Montreal would result in a positive association, a correlation coefficient as strong as 0.79, which is also statistically significant. This would suggest that the positive association between ethnic diversity and trust level in Canadian cities is likely to demonstrate a unique Canadian trend, different from those observed in other countries, an interesting subject that we will revisit in the coming chapters.

Table 7.1: Correlation of trust and ethnic/cultural diversity

	Direction	Pearson Correlation	Sig
Canada	+	0.75	0.00
Canada minus Quebec	+	0.17	0.33
Quebec Only	-	-0.13	0.75
Quebec minus Montreal	+	0.34	0.46
Canada minus Montreal	+	0.79	0.00

The last hypothesis regarding the predictors of a city's trust level revolved around the impact of the size of its immigrant population. As mentioned earlier, the existing literature suggests a negative relationship between the two; that is, the greater the population of immigrants in a city, the lower its overall level of trust. Reporting these for Canadian cities, Figure 7.14 indicates that cities with a small immigrant population tend also to have lower levels of trust, and with the increase of the former the latter also rises. This positive association is reflected in the barely significant yet positive value of the correlation coefficient for the two variables (reported in Appendix 7.2). This, also, is contrary to the findings of the studies in other immigrant-receiving countries, particularly the European ones.

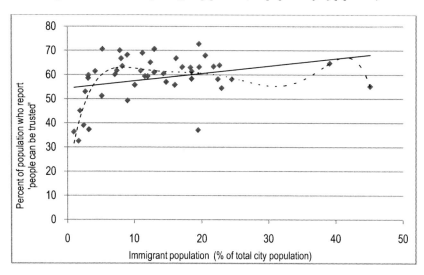

Figure 7.14: Trust Level by immigrant population (as a proportion of city population)

A closer examination of the above graph points to the possibility that the nature of the relationship between the two variables is not necessarily linear (the dotted line in Figure 7.14). As it is evident from the trend line, with the initial increase in the proportion of the immigrant population, the trust level also increases; however, after that initial increase the curve flattens and remains so. Both these features – i.e., the positive linear, and the curvilinear relationship – point to interesting contrasts between Canada and other industrial countries, and open an area for further research.

Possible Compounded Effects?

A potential problem with our discussion so far is that all the factors studied have been examined in isolation from each other. Although this is a useful, and necessary, first step, it is far from refined, as there exists the potential for overlap and interactions between many of these variables. It is through a simultaneous examination of these variables that we can arrive at a more accurate picture showing the pure effect of each. This has been done using a regression model of the predictors of trust, the results of which are reported in Table 7.2.

Table 7.2: Regression Model of Predictors of Trust

	Unstandardized Coefficients		Standardized Coefficients	t	Sig
	B	Std. Error	Beta		
Population	-0.01	0.00	-0.18	-3.65	0.00
Immigrant population (%)	-0.30	0.20	-0.08	-1.49	0.14
Immigrant population (count)	0.02	0.01	0.12	2.56	0.01
Average income	0.00	0.00	0.40	3.82	0.00
Ethnic Diversity (IQV)	43.66	6.75	0.58	6.47	0.00
Incidence of low income in 2000 - Percent (Economic Families)	0.69	0.35	0.14	1.99	0.05

The regression results add a number of elements to the picture acquired above. First, the effects of all the discussed variables on trust are statistically significant except for immigrant population as a percentage of the city's population and poverty rate (whose significance value is just above the acceptable threshold of 0.05). Second, out of all the variables included, ethnic diversity has remained the most influential factor in its impact on the overall level of trust in the city. Third, the nature of the relationship between trust and each of the variables has remained the same as what we discussed earlier based on the bivariate graphs, with two exceptions: immigrant population and poverty rate; the former changed direction from positive to negative, and the latter from negative to positive.

The last point above poses a difficult question: why would the simultaneous inclusion of immigrant population and poverty rate cause such a drastic shift in the behaviours of these two variables? In other words, why has the negative effect of poverty on the overall level of trust among a city's residents, which appeared so consistently in Figures 7.5 through 7.7, changed into a positive impact once the other variables were factored in? The limitations of the existing data do not allow us to address this question rigorously. However, a possibility with some tentative empirical backup can be raised here, and that has to do with the relationship between immigrants and poverty.

Figure 7.15 contains the poverty rates of immigrants and non-immigrants in major Canadian cities. It shows that the immigrant sub-population in each city tend to include a disproportionately higher number of people living below the poverty line. This might mean that the poverty of the city is disproportionately absorbed by its immigrant population, leaving the non-immigrant sub-population in relatively better shape in terms of their income status. This better economic status can in turn result in a higher likelihood of trusting others. Given that the native-born constitute the majority of the population in every single city, their higher level of trust would inevitably result in a higher overall level of trust in the city.

This theoretical scenario has many components, and confirming its validity depends heavily on the availability of empirically-based supporting information for each of those components. One such component is whether or not those who generally have trust in others will change their answers when asked about trust in specific groups of people, such as immigrants, racial minorities, and so forth. If, for instance, such people have lower trust towards immigrants, and also towards those in low-income statuses, the combination of an immigrant and poor population will absorb most of the distrusting heat, so to speak, and leave the level of trust in the rest of society intact.

Figure 7.15: Poverty rate by immigrant status, Canadian CMAs, 2001

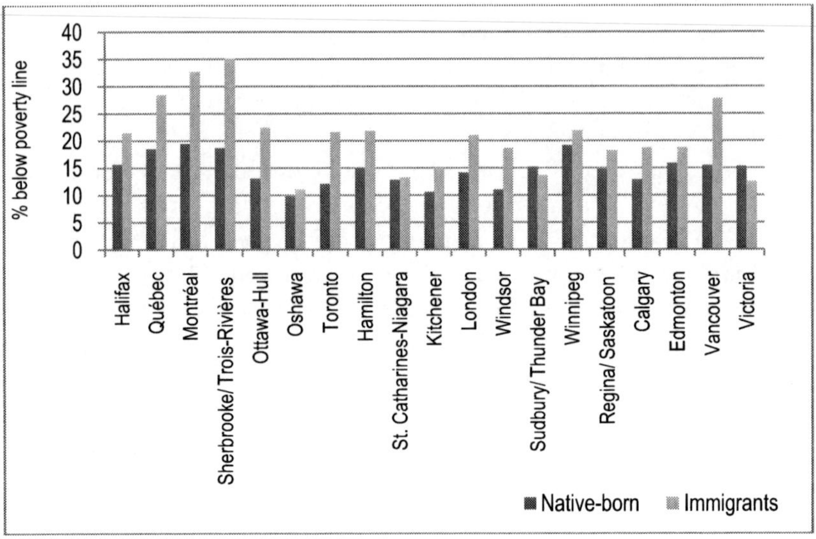

Source: 2001 Census PUMF (Individual)

Conclusion

Before proceeding to the next chapter, it would help to have a re-cap of the major findings of the current chapter. In a nutshell, the level of trust in Canadian cities seems to be strongly but negatively influenced by the size of the city's population, and more strongly but positively by its degree of ethnic diversity. Also, the average income, poverty rate, and percentage of the city's population who are immigrants also correlate with the city's level of trust, although to a lesser degree.

Two of the findings of this study are particularly noteworthy and warrant further attention in future research. One is the relationship between ethnic diversity and trust. The positive relationship found here runs contrary to many other existing studies, particularly those conducted in the United States and Europe. We need further studies before we can argue that this is a uniquely and genuinely Canadian phenomenon, and that it constitutes an exception to the rule established in the literature. Should this happen to be the case, a second question would be what factors have been at work in shaping this so-called *Canadian exceptionalism*.

As a tentative answer to the above question, two factors come to mind: the higher degree of ethnic diversity in Canada, and the positive emphasis put on this diversity by celebrating a multicultural Canada. Social psychological research has long shown that with increased contacts between people of different backgrounds, the existing stereotypes start to shatter and a more trustful relationship begins to appear (see, for instance, Sherif et al, 1961; Aronson, 1992; Isajiw, 1999). A larger exposure of Canadians to people of different ethnic backgrounds might have worked to reduce their fear of 'strangers' and, through that, to raise their level of trust in unknown others. The celebration of multiculturalism may also have intensified such dynamics.

A second interesting finding involves the unique status of Montreal, as a city with a combination of low trust and high diversity. What can explain this anomaly that the ethnic diversity of Montreal has not translated itself into higher trust? We come back to this question later, after discussing the immigration and ethnicity factors in the next few chapters.

8 Trust and Immigrant Status

Trust and international migration are closely related. The reader remembers from the previous chapter that *generalized trust*, the most important type of trust, refers to trust in unknown others or strangers. International migration, indeed, brings strangers together and puts them side by side with each other. So, a true test of trust in a society is perhaps the trust expressed towards and by immigrants.

Recent patterns of international migration give even more salience to this relationship. More and more immigrants are now coming to Canada from countries in Asia, Africa, and the Middle-East that were little known by Canadians, at least compared to European immigrants. They are also coming from cultural environments distant from those of Canadians. In a sense, one may argue that, as a result of such migratory patterns, both immigrants and host populations have become even more 'unknown' to each other. Against such a background, the examination of tendencies to trust among both groups has the potential to reveal a lot about not only immigration but also about trust dynamics.

The issue of trust and immigration is still too new a topic to have generated a rich literature on which we can draw. However, we can still benefit from the existing literature on the factors influencing other aspects of the lives of immigrants – e.g., their economic performance, educational attainment, etc. – to develop a tentative list of important factors that might help with the understanding of how trust and immigration are related. Some such variables are: the economic conditions of the country at the time of their arrival (period of immigration), the degree of immigrants' assimilation into the culture of the host society, and the nature of their social interactions in their new homes. The possible ways in which these factors might influence trust will be discussed in more detail below.

In a previous study of the economic experiences of immigrants, the author found the period of immigration to be particularly important (see, Kazemipur, 2000; Kazemipur and Halli, 2001). This is to say that the economic environment of the country when the immigrants arrive has the potential to leave a long-lasting mark on everything else that would happen to them at a later time. One indication of this importance is the existence of drastically different economic trajectories for immigrants who have arrived in Canada in different decades. Those first few years in the host society act like the first impression in an interpersonal relationship. As the positive or negative nature of those impressions can have a powerful influence on the nature of the relationships between individuals for years to come, the economic experiences of immigrants in the first few years after arrival can also shape their views and perceptions about the host society. It would, therefore, not be too unrealistic to expect that the immigrants' tendencies to trust may also be influenced by their economic experiences in the first few years after their arrival.

Another factor influencing immigrants' experiences in host countries is the degree of their assimilation into the mainstream culture. The assimilation thesis refers to a process through which immigrants pick up and absorb the norms of life in the host society. When the assimilation process is complete, according to this perspective, the outlook of immigrants becomes virtually indistinguishable from those of their native-born counterparts. If, as it was mentioned in the previous chapter, cultural outlooks have a strong influence on people's tendency to trust (or, distrust, for that matter), the gradual shift in the cultural outlooks of immigrants after migration would have clear implications for the level of trust they show towards others. While assimilation *per se* is a difficult concept to measure empirically, two of its correlates easily lend themselves to measurement: generation of immigrants, and their age at arrival. An immigrant who has arrived at a younger age has been more widely exposed to Canadian culture and less to his or her ancestral culture; hence, he or she is more likely to have developed a higher degree of assimilation into the Canadian culture. Similarly, an immigrant whose parents and grandparents have also lived in Canada for part or all of their lives is more likely to develop a Canadian cultural outlook than someone who has just arrived and his or her ancestors are still living in their home country.

Social experiences of immigrants in their new homes are also very influential in shaping their sense of whether they are accepted and welcome in the host society, or excluded and unwelcome. The sense of exclusion that some immigrants might develop as a result of their encounter with the mainstream population goes a long way in creating a sense of belonging to the host society. Here again, the consequences of such feelings can be found in many areas of life, but perhaps more visibly in immigrants' tendencies to trust or distrust others. Against the above background, we will examine some basic facts about the dynamics of trust among immigrants.

Period of Migration

Figure 8.1: Trust, by period of immigration, 1946–2001

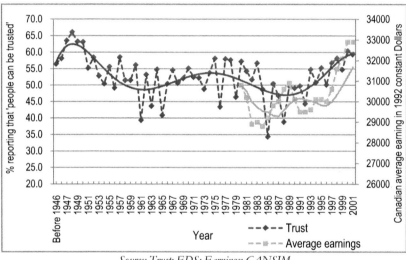

Source: Trust: EDS; Earnings: CANSIM

Figure 8.1 shows two things simultaneously: the proportion of immigrants reporting trust in others by their years of arrival, and the average earnings of Canadians as a whole over time. Both curves show a great deal of fluctuation from one year to another, but the overall shape of the trend lines seem to generally support the following observations. First, the highest levels of trust are reported by those who arrived in Canada immediately after WWII. Given that most such immigrants came from war-trodden

Europe, and had most likely experienced upward mobility in Canada, it is not difficult to see why they had developed such trustful views towards people in general. This upward mobility, or relative improvement, is probably a better predictor of trust. In other words, the key may be the contrast between the economic experiences in the home and host societies, rather than their absolute economic conditions.

Second, for immigrants arriving after that initial post-war period, the trust levels show a decline. The declining trends continue for most of the 1950s and 1960s. Given the absence of any additional data, we can only speculate about the reasons behind those declining trends. Along the lines with the point mentioned above, one possibility is that during the 1950s, many western European countries – the traditional sources of immigrants to Canada, had started recovering from the post-war devastation and were experiencing a more or less similar upward economic movement. This would have diminished the contrast between the experiences of immigrants in their new homes and what they had left behind. The undesirability of migration to Canada from such countries eventually revealed itself through a significant and consistent drop in the number of immigrants arriving from those countries, a pretext for the sweeping immigration policy reforms in Canada during the 1960s, which opened the gates for immigrants from developing societies to start arriving in Canada in large numbers.

Third, the rising level of trust among immigrants during the second half of the 1960s and throughout almost the whole 1970s corresponds also with an increasing number of arrivals from the Third World countries. For such immigrants, the contrast between their old and new homes was significant, much more visible than those for European immigrants of the previous decade, and the chances of upward mobility more conspicuous. These more positive experiences could translate themselves into more pleasant experiences for immigrants, hence, the development of more trusting views.

Fourth, the connection between economic experiences and trust can be more clearly seen for the period 1980–2001, for which some additional data are available. If we take average earnings as a sign of the overall performance of the economy, there seems to be a correspondence between the overall economic conditions of each year and the trust levels of immigrants who had arrived in that year. With some fluctuations, and some lags, the trust levels seem to be lower for those who arrived during the economic recession of the early 1980s, and again for those arriving during the recession of the early 1990s. Since the mid-1990s, both indicators start rising.

Age at Migration

The age at which immigrants arrive in Canada has been shown to be a crucial factor in shaping their later experiences in their new homes. One would expect to see those who arrive at a younger age to be more at home in Canada, given their opportunities to develop better language skills, to become more familiar with local culture, and to possess larger social networks. All of these factors would result in higher social capital in general, and more trust in others in particular. In sum, with the increase of age at arrival, we should see a decline in the trust level.

Figure 8.2 shows the immigrants' level of generalized trust by their age at arrival, allowing for a broad test of the above hypothesis. Although this information was available for each particular age, the years of age were collapsed into broader categories to make the patterns easier to detect. The information presented can be taken to illustrate the following three points; a) contrary to the above expectation of a linear decline of trust with increase in the age of migration, the highest levels of trust are reported by those who arrived at age 50 or more; b) the next highest levels of trust are reported for those arriving under the age of 14; and, c) the lowest levels of trust seem to be reported for those who arrived at between the ages of 15 to 50. The pattern illustrated can highlight a couple of important points suggested in the previous literature on immigration.

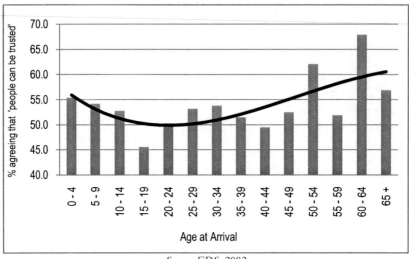

Figure 8.2: Trust level by age at arrival

Source: EDS, 2002

The first point involves the hypothesis suggested in our discussion of the impact of period of immigration, that is, the significant influence of economic experiences on immigrants' tendencies to trust or distrust. As is easily noticeable, one main difference between the high-trust and the low-trust groups – younger and older groups versus the middle-aged one – is that the latter are in their working age and most likely employed, while the former are dependents and either unemployed or with no major obligation to work. The centrality of work naturally leads this group towards a heavier engagement in the Canadian economy and more exposure to the job market; in the process, then, their experiences in this arena become a stronger force in shaping their attitudes towards the general population. Difficulties in the job market, in other words, may have been the reason for the lower level of trust reported by this group. The further decline in trust levels as age increases can also be a product of the increased difficulties faced by those who enter the Canadian job market in the second half of their working years, hence, the lower trust levels reported by those arriving at age 40–45.

For those entering the Canadian job market in their 40s, it is a reasonable expectation that most of the skills they possess have been acquired in their home countries, and are not directly and immediately transferable to the new markets. It is now common knowledge that many such people have to adopt jobs for which they are over-qualified, with the hope that after a while they will be able to find jobs in the professions they have been trained for or have experience in. More often than not, however, such a transition does not take place, and the initial jobs heavily shape their occupational path for the future. Regardless of what happens to these immigrants later, even if they manage to find more satisfying jobs later on, the bitterness of their initial experiences stays with them for a long time. An extreme but instructive example of this dynamic is the situation of those professionals – medical doctors, engineers, etc. – who find themselves driving taxi-cabs or delivering pizza in Canada, after their failure to get their credentials recognized. This type of situation is conducive to the state of constant dissatisfaction and suspicion, both favourable grounds for growth of distrust.

The reverse side of the above phenomenon – i.e., lack of serious concerns about job, skills, training, and economic performance – is probably responsible for the higher than expected level of trust among those migrating at an age of 50 or older. Given the structure of the Canadian immigration point system, in which being in the right age bracket (21–49) awards the applicant with a significant number of points, most of those who arrive at an older age come under family class and as dependents of family members who are already present in the country. Many of these are the parents of economic-class immigrants who are established financially, and who do well enough to convince the immigration authorities that they can support their incoming parents. These immigrants, therefore, do not face many of the challenges and difficulties their children have to go through, are in the company of their children and grandchildren, receive significantly better health care than they would have in their home countries, and are more likely to be satisfied with their migration. The pleasantness of their migration experience can translate itself into higher trust in others.

Another extremely low level of trust in Figure 8.2 is one reported by those immigrants who had arrived between the ages of 15 to 19. The score for this group is about 10 percent lower than those who arrived at the age of 0–4, and about 25 percent lower than the highest score (reported by the age group 60–-64). Before offering any explanation for this, it is noteworthy that a similarly unexpected trend was also detected by the author (see Kazemipur and Halli, 2000) in a study of the economic performance of immigrants, in that a surprisingly higher rate of poverty was found for this group compared to the two age groups immediately before and after them. So, there seems to be something unique in the experiences of this particular group of immigrants that results in such anomalies in the outcomes.

A reliable explanation of this phenomenon, of course, needs more in-depth information, but one possibility can be that immigration is particularly harsh for those migrating as adolescents. What can make this experience so harsh for them can be related to the fact that the post-migration adjustment pressures are simply added to an already intense struggle with an identity crisis for those in this age bracket. As it is known, many adolescents at this stage are under the pressures of trying to establish themselves as independent individuals distinct from their parents, but the advantage of the native-born teenagers is that in this struggle they tend to rely on their peers for emotional support and reinforcement. In the case of immigrant teenagers, they seem to have to do this on their own.

There is another way in which immigration can affect the lives of the adolescent immigrants. Naturally, migration exposes them to a new set of cultural codes at a young age. This otherwise positive development, at this age, would result in the eradication of the sanctity of both the home and the host cultural systems, leaving the making of all major decisions to the cost-benefit analysis of the individual. It can also erode the possibility of effective communication with older generations in general and parents in particular, as such communications normally occur within a shared cultural environment. The so-called cultural limbo that those immigrants find themselves in immediately after arrival, and the lack of clarity about the mutual expectations in their social environments, can easily block them from developing long-lasting relationships with others. In this kind of situation failure is frequent, and failure always nurtures distrust. The amount of mental and intellectual energy demanded by all these tasks can certainly cost these young adults in other areas of life, including school performance and skills development.

Generations of Immigrants

As mentioned earlier, age at immigration could be used as a proxy for degree to which an immigrant has been integrated into the host society and assimilated into the host mainstream culture. Another, perhaps stronger, proxy would be the number of an immigrant's previous generations – parents and grandparents – who were born in the host country. The transition from age at immigration, which focuses only on the so-called first-generation immigrants, to immigrant generation expands the horizon and allows us to see assimilation and integration, and their impact on trust, in a much broader time frame.

Viewing immigrants from a generational perspective, many make a quick distinction between 1st and 2nd generation immigrants, with the former referring to those who have come to Canada as adults and have no parents or grandparents who were born and lived in Canada, and the latter to those who were born in Canada from parents who had migrated to the country as adults. By extension, a 3rd generation immigrant refers to those born in Canada to two Canadian-born parents, while their grandparents were born outside Canada and migrated as adults. Finally, 4th generation immigrants are those who were born in Canada to Canadian-born parents and grandparents (Statistics Canada, 2004).

This image, however, is somewhat simplistic, as the generational traits are sometimes far more elaborate than this. How about, for instance, those who were born outside Canada and migrated at a very young age, say, under 5? Technically, they should still be considered 1st generation immigrants, as they were born outside Canada; but, given their young age at the time of arrival makes them very similar to someone born in Canada, in terms of the degree of their exposure to Canadian culture, their mastery of the official languages, and the type of education they are likely to receive. Another example is someone born in Canada to two parents, only one of whom was born outside Canada. This person would certainly not be a 2nd generation immigrant, but not quite a 3rd generation one, either. To allow for these variations, researchers have expanded the generational perspective so that each of the in-between categories could also have their own labels.

In the Canadian Ethnic Diversity Survey conducted by Statistics Canada, the following generational classification has been used

- 1.0 generation – Born outside Canada, arrived in Canada at age 15 or older
- 1.5 generation – Born outside Canada, arrived in Canada before age 15
- 2.0 generation – Born in Canada with both parents born outside Canada
- 2.5 generation – Born in Canada with one parent born in Canada, one parent born outside Canada
    - born in Canada, 1 parent born in Canada, 1 parent and 4 grandparents born outside Canada
    - Born in Canada, 1 parent and 1 grandparent born in Canada, 1 parent and 3 grandparents born outside Canada
    - Born in Canada, 1 parent and 2 grandparents born in Canada, 1 parent and 2 grandparents born outside Canada
    - Born in Canada, 1 parent and 3 grandparents born in Canada, 1 parent and 1 grandparent born outside Canada
    - Born in Canada, 1 parent and 4 grandparents born in Canada, 1 parent born outside Canada
- 3rd generation – Born in Canada, both parents born in Canada and at least one grandparent born outside Canada
    - Born in Canada, 2 parents born in Canada, 4 grandparents born outside Canada
    - Born in Canada, 2 parents and 1 grandparent born in Canada, 3 grandparents born outside Canada
    - Born in Canada, 2 parents and 2 grandparents born in Canada, 2 grandparents born outside Canada
    - Born in Canada, 2 parents and 3 grandparents born in Canada, 1 grandparent born outside Canada
- 4th generation or more – Born in Canada, both parents and all four grandparents born in Canada

As we move through the above categories from top down, we indeed move from categories of immigrants with the shortest amount of time in Canada and smallest amount of involvement in Canadian social life and exposure to Canadian culture, to those with more involvement and exposure. For the purpose of our study, it is incredibly important, and fascinating, to examine the interaction of an immigrant's generational status and his or her tendency to trust others. To reduce the crowdedness of the above classification, we have compared three broad groups of the generational categories below.

Figure 8.3 shows the level of trust reported by three groups of immigrants: those who were born outside Canada and who migrated when aged 15 or older (1st generation immigrants), those born outside Canada and migrated before the age of 15, and those born in Canada but to immigrant parents. Roughly speaking, these three groups correspond to 1st, 1.5, and 2nd generations. The figures show that the highest level of trust is reported for the middle group, and that those for the other two groups are roughly the same.

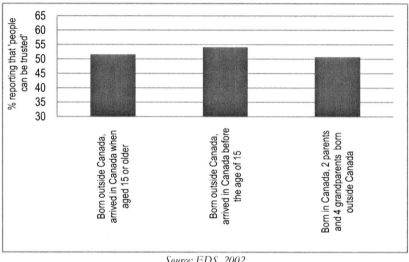

Figure 8.3: Trust and immigration generation, 1st through 2nd generation

Source: EDS, 2002

A good explanation of these variations is difficult to offer in the absence of more sophisticated data, but here again, one could suggest some hypotheses. Towards this goal, I would like to suggest that the difference between the middle category and the other two is that the people in the former group have spent the formative years of their lives in two social and cultural environments, while the latter have been exposed to only one environment, either in Canada or outside. Such exposure to diversity may have had the impact of making this group more tolerant and receptive of difference, and less prone to judging others through stereotypes, one example of which is distrust. This is a hypothesis that seems relevant for explaining the patterns in the next two figures.

Figure 8.4 compares the trust levels for various categories of those born in Canada, who also have one Canadian-born parent. The difference between the four categories included is in the number of Canadian and foreign-born grandparents. Starting from the left, the first bar shows those with all four grand-parents being foreign born. As we move to the right, the number of Canadian-born grand-parents starts to increase, from one to four. Interestingly enough, the trust level jumps for those with only one Canadian-born grandparent, but as this number continues to increase, the level of trust drops. The increase in the number of Canadian-born grandparents, of course, means a decrease in the number of foreign-born grandparents and, as a result, a decrease in exposure to the diversity that could come from having family roots outside Canada. Here again, it seems that the maximum level of trust is not necessarily found in situations of low diversity.

Figure 8.4: Trust and generational status, for those born in Canada, and with one parent born in Canada

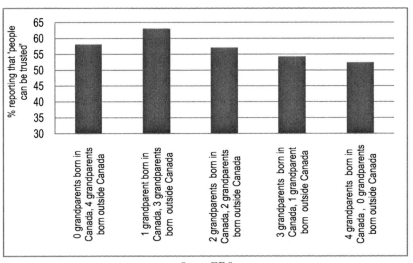

Source: EDS

Interestingly enough, the exact same pattern emerges in Figure 8.5, which shows the level of trust for those born in Canada to two Canadian-born parents. Here again, the maximum level of trust is reported when there is a combination of one Canadian-born and three foreign-born grandparents, and as the number of the latter declines, the level of trust also drops, consistently.

Figure 8.5: Trust and generational status, for those born in Canada, with their two parents born in Canada

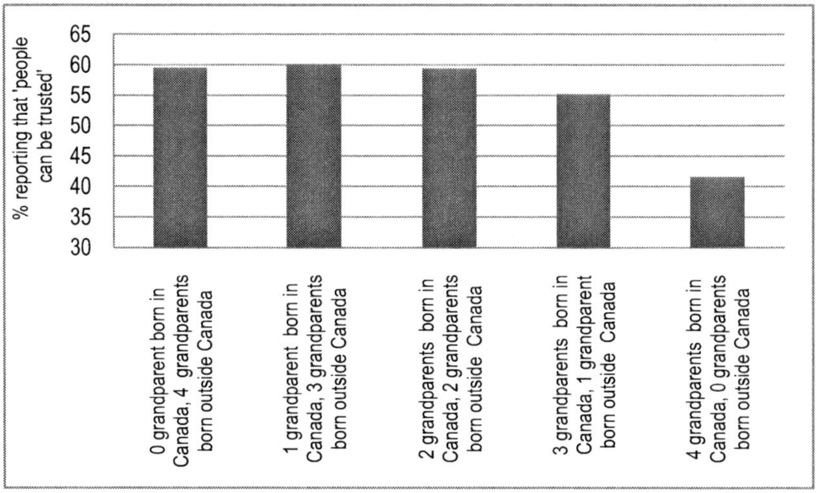

[Bar chart showing % reporting that 'people can be trusted' by grandparent birthplace composition:
- 0 grandparent born in Canada, 4 grandparents born outside Canada: ~59
- 1 grandparent born in Canada, 3 grandparents born outside Canada: ~59
- 2 grandparents born in Canada, 2 grandparents born outside Canada: ~59
- 3 grandparents born in Canada, 1 grandparent born outside Canada: ~55
- 4 grandparents born in Canada, 0 grandparents born outside Canada: ~42]

Source: EDS

The above patterns can be taken to suggest that a higher level of exposure to diversity is associated with a more trustful view towards the population at large. Such an exposure to diversity can materialize as a result of migration at a young age due to constant interaction with family members who are from different national, ethnic, and cultural backgrounds. As the composition of ancestral birthplaces become more homogeneous – that is, they are all born in Canada, or they are all born outside Canada – the level of trust seems to be dropping.

If the above argument could be verified with further research and more comprehensive data, it would point to an extremely important and a very interesting theoretical finding. The conceptual core of such a finding is that, while differences in the substances of different cultures can certainly influence one's tendency to trust, the mere exposure to more than one culture may be a much stronger factor which could overwrite those initial differences. As far as trust is concerned, having been exposed to, and influenced by, more than one culture seems to make it easier for people to trust the anonymous other.

How can one's ancestral cultural diversity create a more favourable ground for trusting others? The mechanism through which the former influences the latter seems to be similar to what we found in chapter 6 about the positive relationship between a city's ethnic diversity and its trust level. As mentioned there, one's exposure to a multitude of cultures has the potential to weaken the negative stereotypes that might hinder the development of more trustful views towards others. It can, in other words, introduce a certain degree of flexibility and acceptance of others in one's eyes. We will come back to this discussion in the final chapter.

Conclusion

The discussion on the interaction of trust and immigration status points to a number of important dynamics. First, the trust levels of immigrants seem to be influenced, to a large extent, by their initial experiences at the time of arrival; those landing at the time of economic booms tend to have a generally higher level of trust, and those with a rough start have a lower level. Second, a higher exposure of immigrants and native-born Canadians to cultural and ethnic diversity tends to be associated with a stronger tendency to trust others. This exposure seems to add a certain degree of plasticity to one's attitude towards others, making it easier to trust those who are completely familiar to them.

Hearing the latter part of the previous paragraph, a student of sociology would probably be reminded of the term used by Emile Durkheim: *the state of plasticity*. His argument from more than a century ago has some hints for what we are concerned with here. In his discussion on the positive functions of crime for society, Durkheim argued that crime poses a challenge to the current collective sentiments and, consequently, prevents them from becoming so rigid that they lose their capacity to change. His words are worth quoting at some length:

> ... for these transformations [of law and morality] to be made possible, the collective sentiments at the basis of morality should not prove unyielding to change, and consequently should be only moderately intense. If they were too strong, they would no longer be malleable... The more strongly a structure is articulated, the more it resists modification... If there were no crimes, this condition would not be fulfilled...Where crime exists, collective sentiments are not only in the *state of plasticity* necessary to assume a new form, but sometimes it even contributes to determining beforehand the shape they will take on (Durkheim, 1982: 101–102. Italics added).

To rephrase Durkheim, one may argue that crime – or deviance, to use a more sociological term – is, indeed, nothing but a life-style different from that of the mainstream population. The presence of deviance exposes people in the mainstream to things they are not used to, and behaviours that are distant from their own. When this distance becomes too noticeable, they may react through sanctioning the deviant. But before the deviant behaviour reaches that extreme point, it is tolerated. Both of these can lead to more awareness of difference; they can also nurture a more tolerant view towards others.

The essence of the above argument is directly applicable to the relationship between exposure to diversity and trust. A different lifestyle and cultural outlook can function, in the eyes of the mainstream population, as tolerable deviance. It exposes them to difference, and can raise their tolerance for difference. Enough exposure to such cultural differences has the potential to broaden the circle of 'normal' behaviour and, hence, of 'acceptable' people. Culturally 'acceptable' people are then more likely to be trusted. Without such an exposure to diversity, and without the resultant plasticity, any minor deviation from one's cultural comfort zone could be considered deviant and punishable, pushing the doer outside the circle of trust.

9 Trust and Ethnicity: Culture

The previous chapter showed that the level of trust reported by immigrants is heavily influenced by several different facets of their experiences as immigrants, that is, the timing of their arrival, their age at migration, their immigration generational status, and the diversity in their ancestral background. While useful for isolating the impacts of international migration, such a discussion may leave the inaccurate assumption that immigrants constitute a homogeneous group and should therefore be treated as such. The rich research on immigration over the past couple of decades has clearly demonstrated the enormous amount of variation that exists within the immigrant sub-population. One source of such variation is the differences between immigrants in terms of their ethnic/cultural backgrounds, which is also closely related to their national origins. In this chapter, we examine this aspect of diversity and its interaction with trust in a Canadian context.

As in the previous chapters, it would be useful to start with the big picture to see whether or not ethnic/cultural background matters at all. Figure 9.1 shows the proportion of people of similar ethnic/cultural background who trust others in general. The graph shows a sizeable difference – more than 60 percentage points – between the highest- and the lowest-trusting groups, indicating that ethnicity does indeed matter, as far as trust is concerned.

The mere presence of these differences and their sheer sizes do not, of course, say much about the possible forces that have created those differences. To what extent can we attribute such differences, or at least a portion of them, to ethnicity itself? This is a relevant hypothesis, given the studies that have shown the influence of early socialization and the overall cultural orientations in shaping one's tendency to trust or distrust others (see, for instance, Offe, 1999). One can certainly argue that such orientations are themselves products of other factors such as historical experiences and social and economic statuses. There is a great deal of validity in those arguments, however, they do not negate the fact that the

Figure 9.1: Trust by ethnicity, Canada, 2002

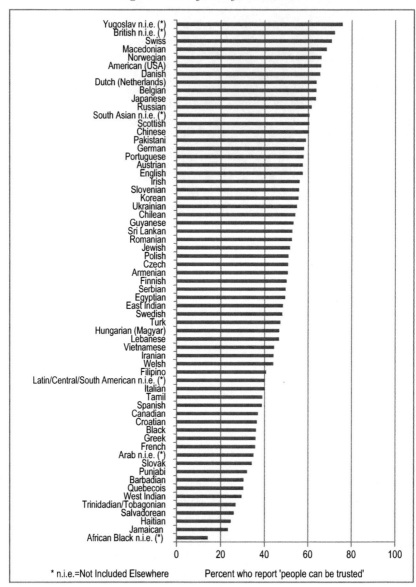

differences among ethnic groups exist, even if they are the embodiment of some larger social and historical forces. Our goal here is to examine the extent to which such differences are real and not proxies for other immediate factors, at least not entirely.

To directly examine the impact of ethnicity on trust is a difficult, if not an impossible, task. However, there are two indirect ways through which we can isolate this impact. One is through comparing the possible differences between immigrants and non-immigrants of similar ethnic origin. Another is a comparison between the trust levels of each ethnic group with that of the country in which that ethnic group constitutes the majority. The presence of a strong similarity between the trust levels in each case can be taken as showing the lasting impact of ethnic origin, an impact that has not been undermined by different immigration statuses and different countries of residence.

Figure 9.2 shows the trust levels reported by immigrants and non-immigrants of similar ethnic origin. The trend line in Figure 9.2 is a rising one, indicating a positive association between the scores reported for immigrant and non-immigrant members of each ethnic group. In a very general way, this can be taken to suggest that the ethnic origin exerts a heavier influence on tendencies to trust than immigration status does. As with any general trend, of course, there are groups that do not cluster around the trend line, including the French, East Indian, Portuguese, Korean, and English, for whom immigrants have reported a higher level of trust in people, and the Austrian and Armenian, for whom the native-born have reported a higher value. No doubt, for all these cases, there is something about the immigration experience that sets immigrants apart from their native-born co-ethnics, but these do not undermine the predominant trend which shows consistency of the trust level reported by all who belong to a certain ethnic group.

Another, perhaps stronger, piece of evidence supporting the significance of ethnicity comes from comparing the trust level reported by each of these groups with that of their corresponding home countries. If culture and ethnicity matter, we should expect to see a low level of trust in countries whose immigrants to Canada have also reported a low value, and vice versa. Figure 9.3 shows the relationship between the trust level of each ethnic group and that of the nations those groups correspond with. The rising trend line again indicates that there is a positive asso-

Figure 9.2: Trust, by ethnicity and immigration status, Canada, 2002

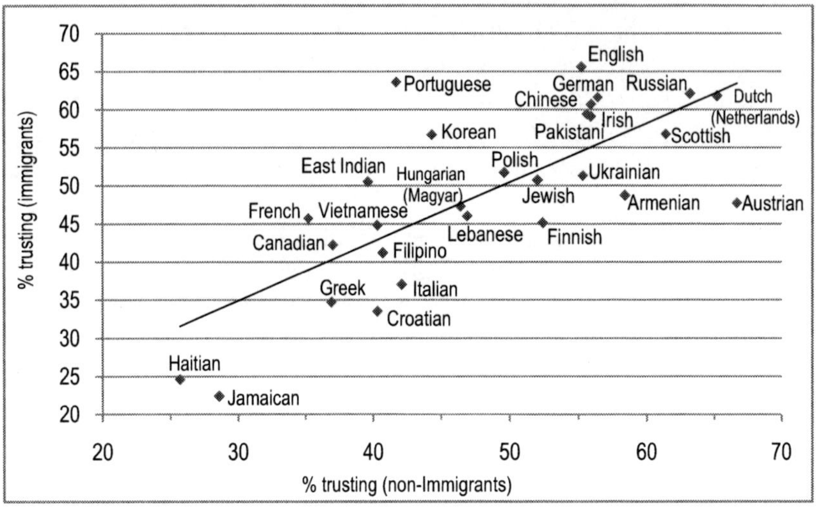

[Scatter plot with x-axis "% trusting (non-Immigrants)" ranging from 20 to 70, and y-axis "% trusting (immigrants)" ranging from 20 to 70. Data points labeled: English, Portuguese, German, Russian, Chinese, Dutch (Netherlands), Irish, Korean, Pakistani, Scottish, East Indian, Hungarian (Magyar), Polish, Ukrainian, French, Vietnamese, Jewish, Armenian, Austrian, Canadian, Filipino, Lebanese, Finnish, Greek, Italian, Croatian, Haitian, Jamaican.]

Source: EDS, 2002

ciation between the two, that is, the groups coming from a country a with higher trust level tend also to trust others more while in Canada. Of course, the number of groups included in this graph is fewer than those reported above, due to the fact that the matching national data were not available for all of them; however, the cases included confirm that, for instance, the groups coming from Nordic countries do report a higher level of trust, compatible with the higher scores of their countries of origin. On the other hand, the ethnic groups coming from former socialist nations or those run by tyrannies report a lower level of trust, again compatible with those reported by the citizens of their home countries.

We should note, however, that under the general trend observed in the graph, there exists a great deal of variation. This is certainly expected, as those migrating from a given country are not necessarily typical citizens of their societies. A selection process is normally involved in international migration; in some cases, immigrants are more educated than their average countrymen, which can result in a higher trust level reported by them; in others, they might consist of minorities fleeing persecution and intolerance, carrying a baggage of suspicion and distrust towards others. It is these types of variations that preclude all dots in Figure 9.3 to fall on the trend line.

Figure 9.3: Trust by ethnicity and country of origin

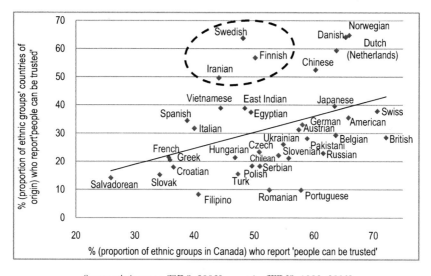

Source: ethnic groups [EDS, 2002], countries [WVS, 1999–2003]

The angle of the slope of the trend line in Figure 9.3 indicates that the association between the trust levels of ethnic groups in Canada and those of their corresponding countries, while positive, is less than perfect. For most groups, the trust levels reported in Canada are noticeably higher than those reported in the corresponding countries. This, in a sense, is a good sign, meaning that migration has been by and large successful for such immigrants, at least to the extent that trust reflects a general satisfaction with life. For at least three groups at the upper sub-diagonal of the graph – consisting of Finns, Swedes, and Iranians – this pattern does not hold, however, meaning that their reported levels of trust in Canada are lower than those reported in their home countries. This is a particularly noteworthy deviation, which needs further attention in the future research. While the lower level of trust among the Finns and Swedes in Canada compared to those in the countries of origin can be referred to the already high levels of trust in those countries, the case of Iranians cannot.

153

Having examined the relationship between trust and culture/ethnicity, we are now in a better position to tackle the issue of the low level of trust in the province of Quebec, which we noticed in chapter five. The reason for the relevance of that discussion is that Quebec is different from the other Canadian provinces, in that the former has a majority population of French ethnic origin. Therefore, given the findings of our chapter so far – that is, that cultural/ethnic backgrounds have a lasting influence on tendencies to trust – it would be reasonable to ask whether the low level of trust in Quebec is also related to the French ethnic origin of its population.

Why a Low Level of Trust in Quebec?

The first step towards addressing the above question is to establish whether or not the low level of trust in the province of Quebec is due to its French-origin population. Figure 9.4 illustrates the overall trust levels for different ethnic groups living in different provinces. Each cluster of bars shows the trust level reported by members of a particular ethnic group, broken down by province of residence. There are two different ways to look into the information provided in this figure. One is to compare clusters with each other, in order to get a sense of whether or not the members of a particular ethnic group have a distinct tendency to trust more or less than other groups. Another way is to close up on each cluster, in order to see whether or not a particular province has a distinct tendency to trust more or less than other provinces, controlling for ethnic groups.

Comparing the cluster of bars for French with all other clusters shows that those of French origin are not particularly less trustful than other groups. Sure enough, the French are not among the highest, but they are not among the lowest, either. They demonstrate trust levels that are lower than groups such as the English, German, and Irish, but they are ahead of the Italians, Aboriginal, South Asians, and the Portuguese. Comparing the provinces – that is, comparing the charts within each

Figure 9.4: Trust, by ethnic/cultural origin and province

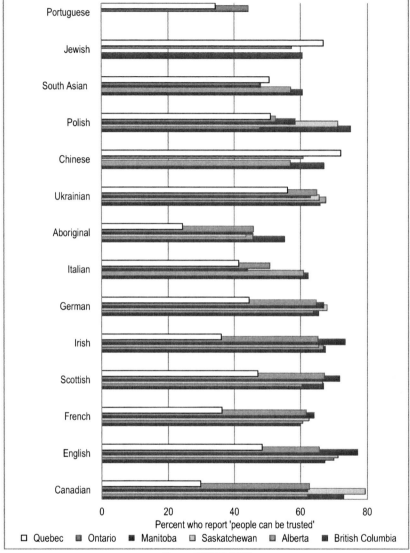

Source: GSS17, 2003

cluster – reveals another aspect of the story. Except for a couple, in almost all clusters, the lowest level of trust is reported for Quebec; that is, people of all ethnic origins report less trust if living in Quebec. The only exceptions are the Chinese, Jewish and, to a lesser extent, South Asians.

This indicates that to find the forces behind the unusually low level of trust in Quebec, we should be looking into the Quebec-related factor rather than ethnicity-related ones. These are the factors that have affected Quebec differently from other provinces.

But, that is not the whole story. Comparing the white bars in all clusters in Figure 9.4 allows for a comparison of French versus other ethnic origins in the province of Quebec. This shows that, within the province, despite the overall tendency to distrust among almost all ethnic groups, the trust level reported by the French is even lower. The only other groups with lower scores are Aboriginal and the 'Canadian.' But, the breakdown of the 'Canadian' group by mother tongue shows that in Quebec, about 96% of the 'Canadians' are primarily French-speaking. So, the problem of low-trust, while a general phenomenon for all ethnic groups in Quebec, is more pronounced for the French-origin or French-speaking segments of population. To that extent, the problem can also be seen as a problem specific to those of French ethnic origin, but only those who live in Quebec and not in the rest of Canada.

Conclusion

The data provided in this chapter showed two things when it comes to people's tendency to trust; that cultural outlook and ethnic heritage matter; and that cultural outlook and ethnic heritage are not the only things that matter. The comparison of immigrants and native-born people of similar ethnic heritage showed a certain level of continuity in terms of trust levels. Also, a comparison of ethnic groups in Canada with the countries whose population's ethnic composition corresponded with those groups also showed a certain level of similarity in trust levels. The examination of trust levels reported by those of French origin in the province of Quebec and in the rest of Canada further corroborated the findings that a certain portion of the trust score reported by each ethnic group can be explained with a reference to their ethnic/cultural tendency.

At the same time, there were variations in the trust levels of the groups being compared that could not be a product of their common ethnic traits. Obviously, like any other social phenomenon, there is a limit in the extent to which a certain variable can explain things. There are always a variety of factors at work simultaneously. One such factor is the role played by the economic variable, to which we turn in the next chapter.

10 Trust and Ethnicity: The Economic Factor

In our discussion in chapter seven, we noticed that cities which have a higher poverty rate tend to report a lower level of trust. The mechanism through which poverty can affect tendencies to trust or distrust is twofold. First, an experience of poverty by a large portion of the population creates among them a sense of alienation towards the rest of the population. As a psychological defence mechanism, people struggling with poverty tend to blame their hardship on the larger society, and subscribe to the view that the larger population has failed them. Such a sentiment can easily translate itself into a distrustful view towards the population at large. Second, living in poverty means living with limited vital resources, and the shortage of such resources would mean that people cannot easily afford to absorb the possible shocks resulting from misplaced trust. Trust in unknown others, as was mentioned earlier, involves a certain degree of risk, and one who trusts therefore needs to be endowed with enough resources to ensure stability in case those risks materialize. Limited resources resulting from poverty can dampen one's tendency and capacity to trust others.

Given the negative relationship we noticed earlier between poverty rates and trust levels in cities, should we expect to find the same thing for specific ethnic groups? Figure 10.1 illustrates the relationship between the poverty rate of each ethnic group and the proportion of them reporting that they trust others. The values reported for the poverty rate and the trust level of ethnic groups are those of their countrywide populations. As expected, the relationship here is also negative, meaning that those groups which experience a higher poverty rate tend to have less trust in others.

The above trend, however, reports the averages at the national level, and such broadly aggregated information can always mask the variations that might exist underneath the surface. In this case, the ethnic communities of similar origin who happen to be living in different cities, for instance, may have different economic experiences. The Chinese living in

Figure 10.1: Trust by poverty rate, ethnic groups, Canada

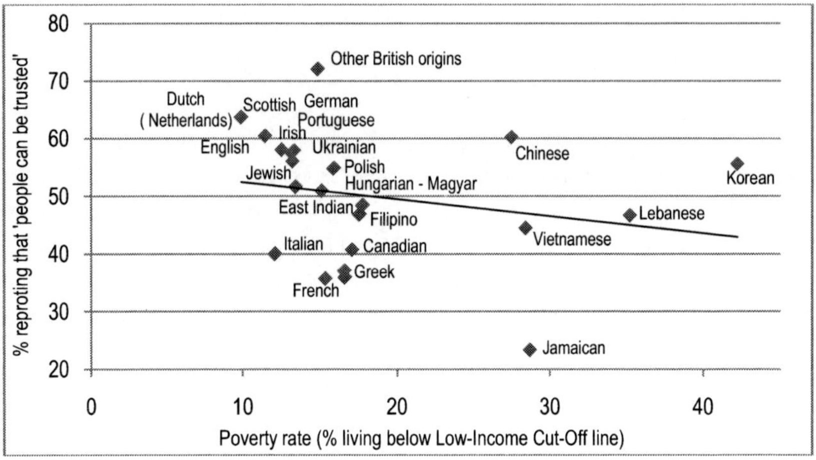

[Scatter plot: x-axis "Poverty rate (% living below Low-Income Cut-Off line)" from 0 to 40; y-axis "% reproting that 'people can be trusted'" from 20 to 80. Data points labeled: Other British origins (~73), Dutch (Netherlands), Scottish, German, Portuguese, Irish, English, Ukrainian, Jewish, Polish, Chinese, Korean, Hungarian-Magyar, East Indian, Filipino, Lebanese, Italian, Canadian, Vietnamese, French, Greek, Jamaican.]

Source: Poverty rate: Canadian census PUMF Individual level, 2001; trust: EDS, 2002

cities with a larger population of co-ethnics, or the French-speaking groups living in a French-speaking environment such as Quebec, for instance, can face a much more favourable and accommodating economic environment compared to their co-ethnics living elsewhere. Furthermore, people live in particular cities and not in the whole country, and the others who are the subjects of their trusting or distrusting views are typically those living in their immediate environments. It would, therefore, make sense to also examine the nature of the association between poverty and trust at smaller geographical levels such as cities and neighbourhoods, which are closer to real life situations.

Figures 10.2 and 10.3 demonstrate the proportion of the population of each ethnic group in each city who trust others, as well as the proportion who live below the poverty line. The cities included are eight of the larger Canadian Census Metropolitan Areas (CMAs) – Toronto, Montreal, Ottawa, Hamilton, Winnipeg, Calgary, Edmonton, and Vancouver – selected because of their larger and more ethnically diverse populations. Each dot on these graphs represents a particular ethnic group in a particular city.

Figure 10.2: The relationship between poverty rate and trust level, larger ethnic groups in 8 larger CMAs

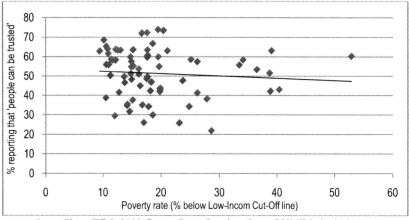

Source: Trust, EDS, 2002; Poverty Rate: Canadian Census-PUMF-Individuals-2001

As Figure 10.2 shows, the relationship between trust and poverty is negative, that is, a high poverty rate of an ethnic group tends to be accompanied by a low level of trust. The slope of the line, which indicates the strength of this association, however, is not very steep. Should this be taken as suggesting that the relationship is weak? As it is the case with scatter plots and the lines representing them, the presence of even a very few outliers can exert a heavy influence on the nature of the line. In this particular graph this seems to be the case, as the right end of the line has moved upward to accommodate about half a dozen cases with extremely high poverty and relatively high trust values. Removing those outliers results in a much more sharply declining line, shown in Figure 10.3, indicating a strong negative relationship between trust and poverty.

The poverty indicator mentioned so far refers to the lack of sufficient income on the part of individuals. This lack of financial resources, as mentioned at the beginning of this chapter, not only deprives one of the resources needed to absorb the possible shocks of misplaced trust but it also nurtures a sense of alienation towards the larger society. There exists, however, another type of deprivation that might have some effects on trust. This other type, which is called 'neighbourhood poverty,' is not directly related to one's individual resources; rather, it refers to the type of neighbourhood in which one lives, in terms of the proportion of its population who live in poverty.

Figure 10.3: The relationship between poverty rate and trust level, larger ethnic groups in 8 larger CMAs (excluding outliers)

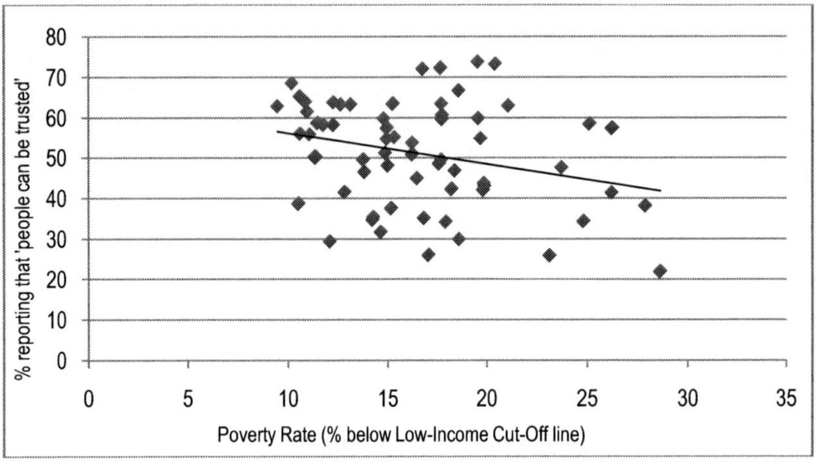

[Scatter plot: x-axis "Poverty Rate (% below Low-Income Cut-Off line)" from 0 to 35; y-axis "% reporting that 'people can be trusted'" from 0 to 80; showing a slight negative trend line.]

Source: Trust, EDS, 2002; Poverty Rate: Canadian Census-PUMF-Individuals-2001

In a previous study, the author has shown that high levels of neighbourhood poverty experienced by some ethnic groups have resulted in lower overall socio-economic performances for those groups (Kazemipur and Halli, 2000a; 2000b). One of the possible consequences of living in high-poverty neighbourhoods – that is, being disproportionately exposed to and surrounded by those living in poverty – has been shown to be the cultural outlooks that people in such neighbourhoods may develop. This type of outlook, which has been called 'the culture of poverty,' affects the way in which one may plan for his or her life primarily through influencing their views on where they stand in relation to others. It is therefore reasonable to expect that exposure to high poverty can affect one's tendency to trust or distrust, as such tendencies are nothing but the way people view others.

Figures 10.4 and 10.5 show the relationship between neighbourhood poverty and trust, for both native-born Canadians and immigrants. Each of the dots in these figures represents a segment of the general population who live in neighbourhoods with the corresponding level of poverty and level of trust. In other words, each dot shows its own particular combination of poverty and trust values, and may therefore represent a small or large group of people. An alternative way to report these two variables was to use each individual as the unit of analysis, and then report the trust level and poverty rates of the neighbourhood in which those individuals live. However, due to the sample size limitations, the trust level reported for each individual neighbourhood would be extremely unreliable. The compromise made here was to combine all neighbourhoods that share the same features, and then treat them as one homogeneous cluster. The dots of the graphs represent such clusters.

The declining trend lines in both figures indicate that as the poverty rate of the neighbourhoods rise, the people living in the neighbourhood become less trustful. There are three important points to keep in mind here. First, that the people who become less trustful may not necessarily live in poverty themselves; it is possible that many have decent financial statuses but, for whatever reason, are living in poor neighbourhoods. A second, and perhaps more important, point is that the increased distrust in such neighbourhoods is directed not necessarily towards other residents of the neighbourhood but towards the population at large. Third, such a negative relationship between neighbourhood poverty and trust is in effect for both the native-born population and immigrants.

Figure 10.4: The relationship between Trust and neighbourhood poverty, native-born Canadians

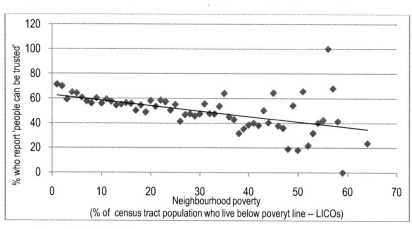

Source: *Neighbourhood poverty: Canadian Census Profile, 2001; Trust: GSS17, 2003*

Figure 10.5: The relationship between Trust and neighbourhood poverty, immigrants

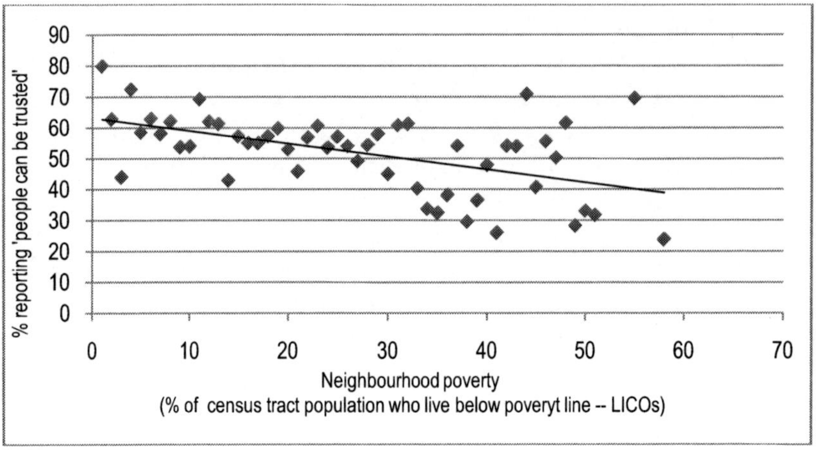

Source: Neighbourhood poverty: Canadian Census Profile, 2001; Trust: GSS17, 2003

Conclusion

The examination of the relationship between poverty and trust shows the strong dampening effect of the former on the latter. It is particularly noteworthy that such an effect holds, regardless of whether we are talking about cities, groups, or neighbourhoods. It was also shown that the declining trust levels are not only a product of the individual experience of poverty but also of the exposure to poverty in the social surrounding. It was hypothesized that such an exposure can trigger cultural forces that would influence the ways in which one views others and their decision whether to trust them or not.

But it is not just the exposure to poverty that can have such an effect. Similar effects can result from some other types of social environments as well, such as the degree to which a particular ethnic group interacts with, or is isolated from, other ethnic groups and the population at large. This is the focus of the next chapter.

11 Trust and Ethnicity: Social Interaction

The differential tendencies of various ethnic groups to trust or distrust others in society are influenced not only by their cultural orientation and economic status, but also by the nature of their social interactions. Indeed, the latter might even be more influential in determining whether 'others' should be trusted or not, because social interaction is the only process that directly involves the 'other.' Cultural outlook and income can act, to a large extent, independently, but social interaction cannot; it requires 'others,' and, incidentally, these 'others' are those whom we would decide to trust or distrust.

Interaction among people takes many different forms, but all of those forms require a favourable environment before they can come into existence. Obviously, some societal spaces, particularly those in which relationships tend to stretch over a longer period of time, are more conducive to social interactions than others. Schools, workplaces, and neighbourhoods, for instance, are some examples of such environments.

In this chapter, we focus on neighbourhoods and their impacts on trust. Previous research has already shown the role played by neighbourhood dynamics on the dwellers' lives (see, for instance, Wilson, 1987; 2000; Fernandez Kelly, 1995; Kazemipur and Halli, 2000). One aspect of the broader dynamics is the important role that neighbourhoods play on dweller's attitudes and prejudices, which has received an incredible amount of attention in the study of racial attitudes among Americans. Given that distrust is, in a sense, a type of prejudicial view towards others – in that, judgements are made about unknown others on the basis of limited information or an over-generalization of some previous observations – the implications of such studies are relevant to our study of trust here.

The studies of the interplay of neighbourhood traits and racial prejudices in America have been greatly informed by the implications of a social-psychological body of research, known as 'contact theory,' one of the few relatively stable and widely supported theses in the social sciences. We will delay a full treatment of this theory until later in this book, but for now it suffices to say that the main thrust of this theory is that the increased interaction between majority and minority groups tends to result in the formation of more positive attitudes among them towards each other. In a study of this phenomenon in Detroit, for instance, Schuman and Hatchett (1974) found that blacks who had reported a higher degree of socialization with their white neighbours had fewer feelings of alienation and distrust towards white society than those who lacked such experiences. Similarly, Sigelman and Welch (1993) have found that such interracial contacts have been associated with "more positive racial attitudes, especially among whites" (p: 781). Along similar lines, Laar et. al. (2005) have reported a similar effect among first-year university students who had roommates of different racial origins; overall, they found that "individuals randomly assigned to live with outgroup roommates at the start of their first year of university showed improved intergroup attitudes by the end of this year" (p: 337). In sum, such interactions seem to help create more balanced and less-prejudicial attitudes, regardless of the context in which they take place.

The mechanism through which such contacts lead to more positive attitudes involves the breakdown of prior prejudices and stereotypes as a result of the friendships made. Pettigrew (1998) lists four interrelated processes that mediate between contact and attitude change: increased knowledge of the previously unknown others, which would shatter negative stereotypes; being in a position to change behaviours towards others, which would in turn push for a change of attitude; generation of affective ties with others not previously considered a part of one's own group; and a resultant reappraisal of one's self and own group, resulting in more appreciation for others' values and less ethnocentric fascination with one's own group.

Neighbourhoods are favourable grounds for facilitating all the above processes. Sigelman et. al. (1996: 1316) argue that people "make friends with others who are available to be friends, and neighbours and coworkers generally fit that description better than anyone else." This, in essence, means that the likelihood of developing trusting views towards others would be seriously limited for those who rarely get exposed to, or interact with, anyone other than people of their own like. Their distrusting views, in this context, are not necessarily a product of some ill intentions on their part; rather, they can be a reflection of a psychological defence mechanism to reduce the natural anxiety resulting from encounters with the unknown.

Against this background, in the following pages we examine the connections between the patterns of spatial distribution of various ethnic groups and the levels of trust reported by them. Given that the study of spatial trends itself is a research area of its own, we start with a discussion of spatial segregation, and then delve into its relationship with trust.

Spatial Segregation: Causes, Consequences, and Patterns

In a general way, spatial segregation of ethnic groups, as conceptualized here, refers to the degree to which the members of each ethnic group are segregated from other groups, in terms of the areas or neighbourhoods of the city in which they live. Hypothetically speaking, in a city composed of only two groups and two neighbourhoods, where all members of one group live in one neighbourhood and all members of the other group in the second one, we have a situation of complete segregation. On the other hand, if the populations of the two groups are evenly distributed in the two neighbourhoods, we have a situation of no or zero segregation. It goes without saying that, with the addition of more groups, the number of possible combinations will also rise, that is, one group may be segregated from a second group but share a neighbourhood with a third.

The causes and the consequences of spatial segregation have been the main concerns of this body of research. Among the potential factors that could shape these differential tendencies are socioeconomic status (i.e., housing affordability), cultural proximity (preference), and discrimination (compulsion). The consequences of residential segregation have also ranged from socioeconomic (e.g., access to job and educational opportunities), cultural (e.g., ability to learn about the mainstream culture and language), and social (e.g., the sense of belonging to, or alienation from, the larger society.) It is in this latter context that we believe that the spatial segregation of ethnic groups can have consequences for their tendencies to trust or distrust. But, before examining the relationship between the two variables, it would be useful to spend some time on spatial segregation itself.

Spatial Segregation of Ethnic Groups: The General Patterns

To quantify the degree of segregation between groups, several statistical measures have been introduced, the most widely used one of which is a measure called the Dissimilarity Index (DI), a measure we have also used in the present study. The reader interested in the history and technical aspects of this measure can consult with Appendix 11.1, but for the purpose of understanding the findings to be presented later, it suffices to know that the DI produces values between 0 and 1, with the former indicating the absence of any segregation between two groups, and the latter, the presence of complete segregation.

In the following few pages, the values of DI will be discussed for the three major cities of Toronto, Vancouver, and Montreal. The choice of these particular cities was made for several reasons. First, it is in these three cities that almost all ethnic groups are present, a situation that allows for a more comprehensive picture of the segregation of each group against all other groups. Second, these ethnic groups have sizeable populations in these three cities, which would make the DI values more reliable and meaningful. Third, a comprehensive examination of all the cities in Canada is unnecessary given that the discussion of the segregation of ethnic groups here is mostly for illustrative purposes. The intention here is primarily to familiarize the readers with the concept so that our discussion of neighbourhood traits and trust in the following section makes more sense.

Against the above background, Table 11.1 contains the values of the DI for all ethnic groups in Toronto. To get an idea of the spatial distribution of one group against all other ethnic groups, one needs to look down each column and compare the reported values. Let's start with

Canadian, that is, those who have picked 'Canadian' as their ethnic/cultural origin. A quick review of the values on the first column of the table shows that 'Canadians' have the least amount of residential segregation with those of western European origins such as the English (0.16), Scottish (0.20), Irish (0.21), German (0.24), French (0.28), and Dutch (0.33). This is then followed by some Eastern European groups like Ukrainian (0.47), Polish (0.47), and Hungarian (0.44). The highest degrees of segregation for the 'Canadians' exist between them and the Quebecois (0.99) and Haitian (0.96), both predominantly French-speaking. Also noteworthy is the fact that out of the groups of Nordic origin, the Swedish and Norwegians have a relatively high level of spatial segregation with 'Canadian' – 0.84 and 0.89, respectively – while the DI values are much smaller for Finnish (0.63) and Danish (0.67).

The second through seventh columns in Table 11.1 report the DI values for the English, French, Scottish, Irish, and German populations. Comparing the values reported for these groups with those of the 'Canadian' group discussed earlier shows a striking similarity between them. In many cases, the two values are either identical or fairly close. This is not surprising, given the fact that all these groups have very small segregation amongst themselves, and that 'Canadian' is a fluid category that sometimes acts as a proxy for one or more ethnic origins; in the case of Toronto, this proximity exists between 'Canadian' and the other six groups listed above, unlike the cities in Quebec in which 'Canadian' is almost identical with French.

The rest of the groups included in the table are those with small populations in the city. Normally, such groups tend to demonstrate higher values of DI, which is a reflection of their small populations, a factor that makes it harder for them to have a noticeable and evenly distributed population in all neighbourhoods across a city. Most such groups consist of recent immigrants coming from small countries. The only exception is the Jewish, a group with a long presence and a noticeably high degree of segregation. This is a common feature of the residential behaviour of Jews in almost every city, which may have its roots in their cultural solidarity, as well as the influence of a history of imposed spatial segregation.

The information included in Figure 11.1, however, can also be approached from a different perspective, in which the focus is on clusters

Table 11.1: Dissimilarity Index, Toronto

	Canadian	English	French	Scottish	Irish	German	Italian	Chinese	Ukrainian	North American Indian	Dutch (Netherlands)	Polish	East Indian	Norwegian	Portuguese	Welsh	Jewish	Russian	Filipino	Métis	Swedish	Hungarian (Magyar)	American (USA)	Greek	Spanish	Jamaican	Danish
Canadian	0.00	0.16	0.28	0.20	0.21	0.24	0.54	0.62	0.47	0.63	0.33	0.47	0.52	0.89	0.57	0.63	0.80	0.68	0.51	0.92	0.84	0.44	0.71	0.52	0.58	0.52	0.67
English	0.16	0.00	0.29	0.17	0.19	0.22	0.56	0.63	0.47	0.65	0.34	0.49	0.57	0.88	0.62	0.62	0.79	0.68	0.54	0.92	0.82	0.44	0.69	0.53	0.61	0.57	0.65
French	0.28	0.29	0.00	0.30	0.30	0.33	0.57	0.63	0.49	0.62	0.42	0.49	0.57	0.89	0.60	0.63	0.80	0.67	0.52	0.91	0.82	0.45	0.71	0.55	0.58	0.57	0.69
Scottish	0.20	0.17	0.30	0.00	0.20	0.26	0.58	0.65	0.49	0.64	0.36	0.51	0.59	0.87	0.63	0.60	0.80	0.69	0.56	0.92	0.82	0.46	0.68	0.54	0.63	0.59	0.65
Irish	0.21	0.19	0.30	0.20	0.00	0.27	0.59	0.64	0.47	0.63	0.37	0.49	0.59	0.88	0.62	0.62	0.80	0.69	0.55	0.92	0.81	0.46	0.68	0.54	0.63	0.59	0.65
German	0.24	0.22	0.33	0.26	0.27	0.00	0.55	0.62	0.46	0.68	0.37	0.47	0.57	0.88	0.62	0.63	0.79	0.66	0.55	0.92	0.82	0.43	0.70	0.53	0.62	0.59	0.67
Italian	0.54	0.56	0.57	0.58	0.59	0.55	0.00	0.68	0.60	0.79	0.64	0.59	0.59	0.92	0.58	0.80	0.83	0.73	0.61	0.95	0.91	0.58	0.82	0.58	0.55	0.58	0.83
Chinese	0.62	0.63	0.63	0.65	0.64	0.62	0.68	0.00	0.68	0.80	0.71	0.69	0.56	0.91	0.71	0.84	0.78	0.72	0.56	0.94	0.89	0.64	0.82	0.53	0.69	0.63	0.85
Ukrainian	0.47	0.47	0.49	0.49	0.47	0.46	0.60	0.68	0.00	0.71	0.58	0.37	0.65	0.88	0.61	0.73	0.80	0.65	0.59	0.94	0.84	0.49	0.74	0.62	0.48	0.48	0.52
North American Indian	0.63	0.65	0.62	0.64	0.63	0.68	0.79	0.80	0.71	0.00	0.70	0.71	0.78	0.88	0.74	0.74	0.88	0.78	0.71	0.90	0.88	0.67	0.76	0.75	0.73	0.73	0.80
Dutch (Netherlands)	0.33	0.34	0.42	0.36	0.37	0.37	0.64	0.71	0.58	0.70	0.00	0.59	0.65	0.87	0.67	0.63	0.83	0.74	0.67	0.92	0.82	0.57	0.73	0.65	0.69	0.67	0.66
Polish	0.47	0.49	0.49	0.51	0.49	0.47	0.59	0.69	0.37	0.71	0.59	0.00	0.58	0.88	0.56	0.73	0.85	0.67	0.53	0.91	0.85	0.50	0.75	0.64	0.60	0.60	0.76
East Indian	0.52	0.57	0.57	0.59	0.59	0.57	0.59	0.56	0.65	0.78	0.65	0.58	0.00	0.91	0.60	0.81	0.87	0.74	0.42	0.94	0.93	0.62	0.85	0.57	0.56	0.37	0.85
Norwegian	0.89	0.88	0.89	0.87	0.88	0.88	0.92	0.91	0.88	0.88	0.87	0.88	0.91	0.00	0.91	0.87	0.90	0.91	0.93	0.94	0.85	0.87	0.87	0.92	0.89	0.92	0.88
Portuguese	0.57	0.62	0.60	0.63	0.62	0.62	0.58	0.71	0.61	0.74	0.67	0.56	0.60	0.91	0.00	0.80	0.89	0.80	0.61	0.94	0.88	0.65	0.84	0.67	0.59	0.60	0.84
Welsh	0.63	0.62	0.63	0.60	0.62	0.63	0.80	0.84	0.73	0.74	0.63	0.73	0.81	0.87	0.80	0.00	0.87	0.82	0.80	0.89	0.81	0.69	0.71	0.77	0.81	0.81	0.69
Jewish	0.80	0.79	0.80	0.80	0.80	0.79	0.83	0.78	0.80	0.88	0.83	0.85	0.87	0.90	0.89	0.87	0.00	0.57	0.79	0.98	0.85	0.70	0.82	0.78	0.84	0.88	0.89
Russian	0.68	0.68	0.67	0.69	0.69	0.66	0.73	0.72	0.65	0.78	0.74	0.67	0.74	0.91	0.80	0.82	0.57	0.00	0.68	0.93	0.85	0.60	0.81	0.69	0.70	0.76	0.82
Filipino	0.51	0.54	0.52	0.56	0.55	0.55	0.61	0.56	0.59	0.71	0.67	0.53	0.42	0.93	0.61	0.80	0.79	0.68	0.00	0.93	0.89	0.54	0.80	0.50	0.55	0.45	0.83
Métis	0.92	0.92	0.91	0.92	0.92	0.92	0.95	0.94	0.94	0.90	0.92	0.91	0.94	0.94	0.94	0.89	0.98	0.93	0.93	0.00	0.92	0.92	0.91	0.92	0.93	0.93	0.91
Swedish	0.84	0.82	0.82	0.82	0.81	0.82	0.91	0.89	0.84	0.88	0.82	0.85	0.93	0.85	0.88	0.81	0.85	0.85	0.89	0.92	0.00	0.79	0.78	0.88	0.90	0.94	0.79
Hungarian (Magyar)	0.44	0.44	0.45	0.46	0.46	0.43	0.58	0.64	0.49	0.67	0.57	0.50	0.62	0.87	0.65	0.69	0.70	0.60	0.54	0.92	0.79	0.00	0.72	0.57	0.60	0.61	0.72
American (USA)	0.71	0.69	0.71	0.68	0.68	0.70	0.82	0.82	0.74	0.76	0.73	0.75	0.85	0.87	0.84	0.71	0.82	0.81	0.80	0.91	0.78	0.72	0.00	0.77	0.81	0.83	0.76
Greek	0.52	0.53	0.55	0.54	0.54	0.53	0.58	0.53	0.62	0.75	0.65	0.64	0.57	0.92	0.67	0.77	0.78	0.69	0.50	0.92	0.88	0.57	0.77	0.00	0.64	0.61	0.81
Spanish	0.58	0.61	0.58	0.63	0.63	0.62	0.55	0.69	0.62	0.73	0.69	0.60	0.56	0.89	0.59	0.81	0.84	0.70	0.55	0.93	0.90	0.60	0.81	0.64	0.00	0.49	0.84
Jamaican	0.52	0.57	0.57	0.59	0.59	0.59	0.58	0.63	0.66	0.73	0.67	0.60	0.37	0.92	0.60	0.81	0.88	0.76	0.45	0.93	0.94	0.61	0.83	0.61	0.49	0.00	0.84
Danish	0.67	0.65	0.69	0.65	0.65	0.67	0.83	0.85	0.74	0.80	0.66	0.76	0.85	0.88	0.84	0.69	0.89	0.82	0.83	0.91	0.79	0.72	0.76	0.81	0.84	0.84	0.00
Vietnamese	0.67	0.71	0.67	0.73	0.72	0.71	0.61	0.69	0.69	0.74	0.78	0.62	0.62	0.93	0.59	0.85	0.90	0.80	0.60	0.95	0.93	0.70	0.83	0.71	0.49	0.56	0.87
British, n.i.e.	0.41	0.39	0.44	0.40	0.41	0.42	0.67	0.70	0.56	0.67	0.48	0.58	0.66	0.85	0.68	0.65	0.82	0.71	0.64	0.92	0.81	0.54	0.70	0.62	0.68	0.67	0.71
Austrian	0.57	0.55	0.58	0.54	0.54	0.55	0.73	0.78	0.65	0.73	0.59	0.67	0.76	0.88	0.78	0.69	0.83	0.75	0.74	0.93	0.79	0.63	0.72	0.70	0.75	0.75	0.72
Lebanese	0.67	0.67	0.67	0.68	0.68	0.66	0.73	0.69	0.70	0.81	0.74	0.64	0.63	0.91	0.74	0.82	0.87	0.77	0.62	0.94	0.90	0.67	0.84	0.66	0.71	0.69	0.81
Romanian	0.61	0.62	0.61	0.62	0.62	0.61	0.68	0.69	0.63	0.76	0.70	0.63	0.69	0.89	0.75	0.78	0.73	0.63	0.60	0.91	0.83	0.54	0.80	0.62	0.67	0.70	0.80
Belgian	0.88	0.87	0.88	0.87	0.88	0.88	0.93	0.91	0.90	0.91	0.85	0.89	0.91	0.85	0.93	0.83	0.92	0.93	0.92	0.92	0.82	0.89	0.84	0.92	0.92	0.92	0.89
Finnish	0.63	0.62	0.62	0.61	0.60	0.61	0.79	0.76	0.72	0.73	0.65	0.75	0.77	0.88	0.83	0.70	0.83	0.75	0.74	0.93	0.81	0.66	0.74	0.71	0.80	0.79	0.71
Swiss	0.78	0.77	0.76	0.76	0.77	0.75	0.88	0.90	0.78	0.83	0.75	0.79	0.87	0.85	0.88	0.76	0.91	0.85	0.88	0.93	0.83	0.80	0.76	0.87	0.88	0.88	0.76
Korean	0.60	0.59	0.59	0.61	0.61	0.57	0.68	0.55	0.57	0.77	0.68	0.60	0.62	0.89	0.72	0.76	0.69	0.57	0.56	0.94	0.84	0.54	0.78	0.61	0.65	0.66	0.79
Québécois	0.99	0.99	0.98	0.99	0.99	0.99	0.99	0.99	0.99	0.98	0.99	0.98	0.99	0.99	0.99	1.00	1.00	0.99	0.99	0.97	0.99	0.98	0.99	0.99	0.99	0.99	1.00
African (Black), n.i.e.	0.66	0.69	0.67	0.71	0.70	0.71	0.72	0.73	0.74	0.73	0.76	0.70	0.59	0.91	0.69	0.84	0.91	0.77	0.59	0.94	0.94	0.70	0.85	0.72	0.57	0.49	0.86
Croatian	0.57	0.57	0.57	0.59	0.58	0.55	0.63	0.75	0.52	0.78	0.63	0.46	0.63	0.89	0.61	0.74	0.89	0.76	0.64	0.94	0.84	0.59	0.79	0.70	0.65	0.67	0.79
Iranian	0.65	0.65	0.66	0.66	0.67	0.64	0.74	0.58	0.70	0.83	0.72	0.71	0.68	0.88	0.81	0.82	0.70	0.61	0.64	0.94	0.85	0.62	0.82	0.61	0.74	0.72	0.83
Japanese	0.54	0.52	0.53	0.52	0.51	0.52	0.69	0.59	0.61	0.74	0.63	0.62	0.65	0.89	0.71	0.72	0.78	0.70	0.59	0.92	0.81	0.56	0.74	0.56	0.69	0.67	0.76
Haitian	0.96	0.96	0.95	0.95	0.95	0.96	0.96	0.94	0.97	0.96	0.96	0.97	0.94	0.92	0.94	0.97	0.95	0.97	0.94	0.95	0.97	0.95	0.95	0.95	0.94	0.93	0.96
Czech	0.65	0.63	0.64	0.63	0.62	0.62	0.77	0.77	0.65	0.76	0.69	0.65	0.76	0.88	0.78	0.73	0.82	0.74	0.72	0.91	0.85	0.64	0.71	0.73	0.76	0.76	0.77
Icelandic	0.97	0.97	0.97	0.96	0.97	0.97	0.97	0.99	0.97	0.97	0.97	0.98	0.98	0.97	0.99	0.96	0.97	0.96	0.98	0.99	0.93	0.95	0.94	0.97	0.98	0.99	0.97
Pakistani	0.65	0.68	0.66	0.70	0.68	0.68	0.70	0.64	0.71	0.77	0.76	0.63	0.46	0.91	0.71	0.83	0.91	0.75	0.54	0.90	0.93	0.70	0.84	0.66	0.64	0.55	0.87
Arab, n.i.e.	0.65	0.68	0.66	0.69	0.68	0.67	0.73	0.65	0.69	0.80	0.73	0.62	0.59	0.92	0.72	0.80	0.89	0.73	0.61	0.94	0.91	0.67	0.83	0.69	0.65	0.61	0.82
Acadian	0.97	0.97	0.96	0.97	0.96	0.97	0.98	0.98	0.95	0.97	0.97	0.95	0.98	0.99	0.98	0.97	0.99	0.98	0.98	0.95	0.97	0.96	0.98	0.98	0.98	0.98	0.96
Yugoslav, n.i.e.	0.63	0.63	0.63	0.64	0.63	0.63	0.72	0.77	0.62	0.75	0.69	0.60	0.72	0.89	0.73	0.76	0.84	0.70	0.66	0.93	0.82	0.61	0.77	0.68	0.69	0.73	0.79
Sri Lankan	0.70	0.73	0.73	0.74	0.75	0.73	0.76	0.63	0.79	0.84	0.82	0.74	0.52	0.94	0.78	0.87	0.94	0.81	0.52	0.94	0.95	0.75	0.88	0.65	0.68	0.53	0.91
West Indian	0.55	0.59	0.59	0.62	0.61	0.61	0.61	0.63	0.69	0.72	0.69	0.63	0.40	0.91	0.63	0.80	0.89	0.77	0.48	0.94	0.93	0.65	0.83	0.61	0.54	0.36	0.81
Inuit	0.98	0.99	0.98	0.99	0.99	0.99	0.99	0.99	0.99	0.97	0.98	0.99	0.98	0.99	0.99	0.98	1.00	1.00	1.00	0.99	0.99	0.99	0.98	0.99	0.99	0.99	0.98
Serbian	0.65	0.64	0.62	0.64	0.63	0.62	0.74	0.75	0.53	0.76	0.70	0.53	0.73	0.90	0.74	0.73	0.83	0.69	0.66	0.92	0.81	0.59	0.78	0.67	0.72	0.76	0.80
Black	0.61	0.65	0.64	0.67	0.66	0.67	0.66	0.70	0.73	0.76	0.74	0.67	0.48	0.91	0.67	0.84	0.90	0.79	0.54	0.93	0.92	0.67	0.85	0.68	0.56	0.39	0.84
Guyanese	0.60	0.64	0.63	0.65	0.66	0.66	0.66	0.67	0.73	0.77	0.72	0.66	0.42	0.92	0.69	0.83	0.92	0.79	0.51	0.94	0.93	0.67	0.86	0.64	0.56	0.38	0.85
Slovak	0.68	0.67	0.67	0.68	0.67	0.66	0.77	0.81	0.63	0.76	0.70	0.64	0.78	0.88	0.75	0.73	0.87	0.78	0.75	0.92	0.83	0.66	0.74	0.77	0.77	0.80	0.78
Trinidadian/Tobagonian	0.55	0.59	0.59	0.60	0.60	0.60	0.64	0.64	0.67	0.70	0.68	0.61	0.50	0.90	0.64	0.77	0.88	0.75	0.49	0.93	0.92	0.61	0.84	0.62	0.55	0.44	0.82
South Asian, n.i.e.	0.64	0.67	0.67	0.69	0.68	0.67	0.70	0.65	0.73	0.80	0.75	0.67	0.39	0.92	0.71	0.84	0.92	0.79	0.51	0.93	0.94	0.70	0.86	0.66	0.63	0.48	0.86
Punjabi	0.80	0.83	0.82	0.84	0.84	0.84	0.78	0.85	0.86	0.88	0.84	0.82	0.55	0.93	0.79	0.90	0.98	0.91	0.75	0.96	0.96	0.85	0.93	0.85	0.77	0.66	0.90
Latin/Central/South American, n.i.e.	0.69	0.71	0.69	0.72	0.71	0.71	0.65	0.77	0.72	0.73	0.78	0.70	0.65	0.93	0.67	0.84	0.90	0.77	0.66	0.93	0.93	0.71	0.84	0.73	0.50	0.58	0.87
Egyptian	0.70	0.71	0.70	0.72	0.71	0.70	0.76	0.61	0.75	0.86	0.77	0.67	0.64	0.92	0.76	0.81	0.86	0.76	0.64	0.95	0.88	0.71	0.84	0.68	0.74	0.67	0.85
Armenian	0.76	0.75	0.77	0.76	0.76	0.74	0.81	0.63	0.81	0.88	0.81	0.82	0.76	0.92	0.87	0.84	0.79	0.76	0.74	0.93	0.88	0.73	0.85	0.69	0.81	0.78	0.85
Average	0.63	0.63	0.64	0.64	0.64	0.64	0.72	0.73	0.69	0.78	0.70	0.67	0.68	0.90	0.74	0.79	0.85	0.76	0.67	0.93	0.87	0.67	0.81	0.70	0.70	0.68	0.81

Source: 2001 Census Profile, CMA - CA - CT level data, Ethnic tabulations - single responses

Vietnamese	British, n.i.e.	Austrian	Lebanese	Romanian	Belgian	Finnish	Swiss	Korean	Québécois	African (Black), n.i.e.	Croatian	Iranian	Japanese	Haitian	Czech	Icelandic	Pakistani	Arab, n.i.e.	Acadian	Yugoslav, n.i.e.	Sri Lankan	West Indian	Inuit	Serbian	Black	Guyanese	Slovak	Trinidadian/Tobagonian	South Asian, n.i.e.	Punjabi	Latin/Central/South American, n.i.e.	Egyptian	Armenian
0.67	0.41	0.57	0.67	0.61	0.88	0.63	0.78	0.60	0.99	0.66	0.57	0.65	0.54	0.96	0.65	0.97	0.65	0.65	0.97	0.63	0.70	0.55	0.98	0.65	0.61	0.60	0.68	0.55	0.64	0.80	0.69	0.70	0.76
0.71	0.39	0.55	0.67	0.62	0.87	0.62	0.77	0.59	0.99	0.69	0.57	0.65	0.52	0.96	0.63	0.97	0.68	0.68	0.97	0.63	0.73	0.59	0.99	0.64	0.65	0.64	0.67	0.59	0.67	0.83	0.71	0.71	0.75
0.67	0.44	0.58	0.67	0.61	0.88	0.62	0.76	0.59	0.98	0.67	0.57	0.66	0.53	0.95	0.64	0.97	0.66	0.66	0.96	0.63	0.73	0.59	0.98	0.62	0.64	0.63	0.67	0.59	0.67	0.82	0.69	0.70	0.77
0.73	0.40	0.54	0.68	0.62	0.87	0.61	0.76	0.61	0.99	0.71	0.59	0.66	0.52	0.95	0.63	0.96	0.70	0.69	0.97	0.64	0.74	0.62	0.99	0.64	0.67	0.65	0.68	0.60	0.69	0.84	0.72	0.72	0.76
0.72	0.41	0.54	0.68	0.62	0.88	0.60	0.77	0.61	0.99	0.70	0.58	0.67	0.51	0.95	0.62	0.97	0.68	0.68	0.96	0.63	0.75	0.61	0.99	0.63	0.66	0.66	0.67	0.60	0.68	0.84	0.71	0.71	0.76
0.71	0.42	0.55	0.66	0.61	0.88	0.61	0.75	0.57	0.99	0.71	0.55	0.64	0.52	0.96	0.62	0.97	0.68	0.67	0.97	0.63	0.73	0.61	0.99	0.62	0.67	0.66	0.66	0.60	0.67	0.84	0.71	0.70	0.74
0.61	0.67	0.73	0.73	0.68	0.93	0.79	0.88	0.68	0.99	0.72	0.63	0.74	0.69	0.96	0.77	0.97	0.70	0.73	0.98	0.72	0.76	0.61	0.99	0.74	0.66	0.66	0.77	0.64	0.70	0.78	0.65	0.76	0.81
0.69	0.70	0.78	0.69	0.69	0.91	0.76	0.90	0.55	0.99	0.73	0.75	0.58	0.59	0.94	0.77	0.99	0.64	0.65	0.98	0.77	0.63	0.63	0.99	0.75	0.70	0.67	0.81	0.64	0.65	0.85	0.77	0.61	0.63
0.50	0.56	0.65	0.70	0.63	0.90	0.72	0.78	0.57	0.99	0.74	0.52	0.70	0.61	0.97	0.65	0.97	0.71	0.69	0.95	0.62	0.79	0.69	0.99	0.53	0.73	0.73	0.63	0.67	0.73	0.86	0.72	0.75	0.81
0.74	0.67	0.73	0.81	0.76	0.91	0.73	0.83	0.77	0.98	0.73	0.78	0.83	0.74	0.96	0.76	0.97	0.77	0.80	0.97	0.75	0.84	0.72	0.97	0.76	0.76	0.77	0.76	0.70	0.80	0.88	0.73	0.86	0.88
0.78	0.48	0.59	0.74	0.70	0.85	0.65	0.75	0.68	0.99	0.76	0.63	0.72	0.63	0.96	0.69	0.97	0.76	0.73	0.97	0.69	0.82	0.69	0.98	0.70	0.74	0.72	0.70	0.68	0.75	0.84	0.78	0.77	0.81
0.62	0.58	0.67	0.64	0.63	0.89	0.75	0.79	0.60	0.98	0.70	0.46	0.71	0.62	0.97	0.65	0.98	0.63	0.62	0.95	0.60	0.74	0.63	0.99	0.53	0.67	0.66	0.64	0.61	0.67	0.82	0.70	0.67	0.82
0.62	0.66	0.76	0.63	0.69	0.91	0.77	0.87	0.62	0.99	0.59	0.63	0.68	0.65	0.94	0.76	0.98	0.46	0.59	0.98	0.72	0.52	0.40	0.98	0.73	0.48	0.42	0.78	0.50	0.39	0.55	0.65	0.64	0.76
0.93	0.85	0.88	0.91	0.89	0.85	0.88	0.85	0.89	0.99	0.91	0.89	0.88	0.89	0.92	0.88	0.97	0.91	0.92	0.99	0.89	0.94	0.91	0.99	0.90	0.91	0.92	0.88	0.90	0.92	0.93	0.93	0.92	0.92
0.59	0.68	0.78	0.74	0.75	0.93	0.83	0.88	0.72	0.99	0.69	0.61	0.81	0.71	0.94	0.78	0.99	0.71	0.72	0.98	0.73	0.78	0.63	0.99	0.74	0.67	0.69	0.75	0.64	0.71	0.79	0.67	0.76	0.87
0.85	0.65	0.69	0.82	0.78	0.83	0.70	0.76	0.76	1.00	0.84	0.74	0.82	0.72	0.97	0.73	0.96	0.83	0.80	0.97	0.76	0.87	0.80	0.98	0.73	0.84	0.83	0.73	0.77	0.84	0.90	0.84	0.81	0.84
0.90	0.82	0.83	0.87	0.73	0.92	0.83	0.91	0.69	1.00	0.91	0.89	0.70	0.78	0.95	0.82	0.97	0.91	0.89	0.99	0.84	0.94	0.89	1.00	0.83	0.90	0.92	0.87	0.88	0.92	0.98	0.90	0.86	0.79
0.80	0.71	0.75	0.77	0.63	0.93	0.75	0.85	0.57	0.99	0.77	0.76	0.61	0.70	0.97	0.74	0.96	0.75	0.73	0.98	0.70	0.81	0.77	1.00	0.69	0.79	0.79	0.78	0.75	0.79	0.91	0.77	0.76	0.76
0.60	0.64	0.74	0.62	0.60	0.92	0.74	0.88	0.56	0.99	0.59	0.64	0.64	0.59	0.94	0.72	0.98	0.54	0.61	0.98	0.66	0.52	0.48	1.00	0.66	0.54	0.51	0.75	0.49	0.51	0.75	0.66	0.64	0.74
0.95	0.92	0.93	0.94	0.91	0.92	0.93	0.93	0.94	0.97	0.94	0.94	0.94	0.92	0.95	0.91	0.99	0.90	0.94	0.95	0.93	0.94	0.94	0.99	0.92	0.93	0.94	0.92	0.93	0.93	0.96	0.93	0.95	0.93
0.93	0.81	0.79	0.90	0.83	0.82	0.81	0.83	0.84	0.99	0.94	0.84	0.85	0.81	0.97	0.85	0.93	0.93	0.91	0.97	0.82	0.95	0.93	0.99	0.81	0.92	0.93	0.83	0.92	0.94	0.96	0.93	0.88	0.88
0.70	0.54	0.63	0.67	0.54	0.89	0.66	0.80	0.54	0.98	0.70	0.59	0.62	0.56	0.95	0.64	0.95	0.70	0.67	0.96	0.61	0.75	0.65	0.99	0.59	0.67	0.67	0.66	0.61	0.70	0.85	0.71	0.71	0.73
0.83	0.70	0.72	0.84	0.80	0.84	0.74	0.76	0.78	0.99	0.85	0.79	0.82	0.74	0.95	0.71	0.94	0.84	0.83	0.98	0.77	0.88	0.83	0.98	0.78	0.85	0.86	0.74	0.84	0.86	0.93	0.84	0.84	0.85
0.71	0.62	0.70	0.66	0.62	0.92	0.71	0.87	0.61	0.99	0.72	0.70	0.61	0.56	0.95	0.73	0.97	0.66	0.69	0.98	0.68	0.65	0.61	0.99	0.67	0.68	0.64	0.77	0.62	0.66	0.85	0.73	0.68	0.69
0.49	0.68	0.75	0.71	0.67	0.92	0.80	0.88	0.65	0.99	0.57	0.65	0.74	0.69	0.94	0.76	0.98	0.64	0.65	0.98	0.69	0.68	0.54	0.99	0.72	0.56	0.56	0.77	0.55	0.63	0.77	0.50	0.74	0.81
0.56	0.67	0.75	0.69	0.70	0.92	0.79	0.88	0.66	0.99	0.49	0.67	0.72	0.67	0.93	0.76	0.99	0.55	0.61	0.98	0.73	0.53	0.36	0.99	0.76	0.39	0.38	0.80	0.44	0.48	0.66	0.58	0.67	0.78
0.87	0.71	0.72	0.81	0.80	0.89	0.71	0.76	0.79	1.00	0.86	0.79	0.83	0.76	0.96	0.77	0.97	0.87	0.82	0.96	0.79	0.91	0.81	0.98	0.80	0.84	0.85	0.78	0.82	0.86	0.90	0.87	0.85	0.85
0.00	0.74	0.81	0.75	0.74	0.93	0.83	0.91	0.73	0.99	0.61	0.68	0.82	0.73	0.94	0.80	0.99	0.65	0.69	0.99	0.74	0.69	0.60	0.99	0.75	0.62	0.61	0.80	0.63	0.66	0.78	0.58	0.77	0.86
0.74	0.00	0.61	0.73	0.67	0.88	0.64	0.75	0.67	0.99	0.73	0.66	0.71	0.58	0.96	0.67	0.97	0.73	0.71	0.97	0.66	0.79	0.67	0.98	0.68	0.73	0.70	0.70	0.65	0.74	0.85	0.74	0.75	0.79
0.81	0.61	0.00	0.76	0.70	0.89	0.67	0.75	0.72	0.99	0.81	0.72	0.75	0.67	0.94	0.67	0.96	0.80	0.75	0.94	0.69	0.80	0.75	0.99	0.70	0.77	0.77	0.71	0.73	0.78	0.87	0.78	0.78	0.83
0.75	0.73	0.76	0.00	0.71	0.91	0.77	0.87	0.66	1.00	0.77	0.69	0.67	0.66	0.97	0.75	0.98	0.68	0.62	0.96	0.70	0.72	0.71	1.00	0.72	0.73	0.72	0.78	0.67	0.68	0.82	0.77	0.62	0.71
0.74	0.67	0.70	0.71	0.00	0.88	0.74	0.85	0.57	0.98	0.76	0.69	0.55	0.65	0.93	0.73	0.96	0.71	0.67	0.96	0.66	0.75	0.72	1.00	0.63	0.74	0.73	0.75	0.70	0.72	0.88	0.76	0.69	0.68
0.93	0.88	0.89	0.91	0.88	0.00	0.89	0.87	0.87	0.97	0.92	0.89	0.89	0.86	0.97	0.90	0.91	0.93	0.89	0.95	0.90	0.93	0.92	0.99	0.89	0.94	0.90	0.87	0.91	0.90	0.94	0.92	0.91	0.92
0.83	0.64	0.67	0.77	0.74	0.89	0.00	0.78	0.71	1.00	0.81	0.79	0.75	0.66	0.96	0.75	0.96	0.81	0.79	0.97	0.75	0.81	0.78	0.99	0.75	0.81	0.80	0.76	0.75	0.80	0.90	0.79	0.81	0.78
0.91	0.75	0.75	0.87	0.85	0.87	0.78	0.00	0.85	0.99	0.89	0.82	0.87	0.80	0.96	0.77	0.96	0.86	0.85	0.95	0.83	0.93	0.88	1.00	0.79	0.89	0.90	0.77	0.85	0.89	0.92	0.91	0.87	0.91
0.73	0.67	0.72	0.66	0.57	0.87	0.78	0.85	0.00	0.99	0.72	0.66	0.49	0.55	0.93	0.71	0.96	0.66	0.64	0.97	0.69	0.72	0.67	0.99	0.63	0.71	0.72	0.71	0.66	0.69	0.86	0.74	0.65	0.69
0.99	0.99	0.99	1.00	0.98	0.97	1.00	0.99	0.99	0.00	0.99	0.99	1.00	0.99	0.97	0.98	1.00	0.99	0.99	0.97	0.99	0.99	0.99	1.00	1.00	0.99	1.00	0.99	0.99	1.00	0.99	0.98	0.99	0.99
0.61	0.73	0.81	0.77	0.76	0.92	0.81	0.89	0.72	0.99	0.00	0.76	0.79	0.74	0.93	0.79	0.98	0.64	0.68	0.98	0.75	0.63	0.55	0.99	0.78	0.54	0.56	0.82	0.56	0.61	0.77	0.62	0.77	0.83
0.68	0.66	0.72	0.69	0.69	0.89	0.79	0.82	0.66	0.99	0.76	0.00	0.76	0.69	0.97	0.70	0.97	0.70	0.66	0.97	0.65	0.80	0.69	0.99	0.61	0.72	0.72	0.67	0.68	0.71	0.80	0.74	0.68	0.85
0.82	0.71	0.75	0.67	0.55	0.89	0.75	0.87	0.49	1.00	0.79	0.76	0.00	0.64	0.90	0.74	0.97	0.71	0.67	0.95	0.75	0.74	0.74	0.99	0.70	0.76	0.76	0.80	0.72	0.73	0.91	0.82	0.63	0.60
0.73	0.58	0.67	0.66	0.65	0.86	0.66	0.80	0.55	0.99	0.74	0.69	0.64	0.00	0.95	0.66	0.97	0.72	0.69	0.96	0.70	0.74	0.66	0.99	0.67	0.72	0.71	0.69	0.65	0.72	0.85	0.74	0.69	0.74
0.94	0.96	0.94	0.97	0.93	0.97	0.96	0.96	0.93	0.97	0.93	0.97	0.90	0.95	0.00	0.95	0.98	0.94	0.96	0.95	0.96	0.93	0.94	1.00	0.96	0.92	0.93	0.96	0.95	0.93	0.92	0.94	0.96	0.94
0.80	0.67	0.67	0.75	0.73	0.90	0.75	0.77	0.71	0.98	0.79	0.70	0.74	0.66	0.95	0.00	0.96	0.78	0.76	0.95	0.71	0.82	0.73	0.99	0.70	0.78	0.76	0.68	0.72	0.79	0.90	0.78	0.78	0.83
0.99	0.97	0.96	0.98	0.96	0.91	0.96	0.96	0.96	1.00	0.98	0.97	0.97	0.97	0.98	0.96	0.00	0.99	0.98	0.97	0.96	0.99	0.99	1.00	0.95	0.99	0.99	0.95	0.99	0.98	0.99	0.98	0.98	0.95
0.65	0.73	0.80	0.68	0.71	0.93	0.81	0.86	0.66	0.99	0.64	0.70	0.71	0.72	0.94	0.78	0.99	0.00	0.63	0.98	0.74	0.57	0.55	0.98	0.72	0.60	0.57	0.79	0.59	0.49	0.68	0.69	0.66	0.79
0.69	0.71	0.75	0.62	0.67	0.89	0.79	0.85	0.64	0.99	0.68	0.66	0.67	0.69	0.96	0.76	0.98	0.63	0.00	0.98	0.72	0.67	0.63	0.99	0.72	0.66	0.64	0.76	0.64	0.61	0.79	0.73	0.60	0.72
0.99	0.97	0.94	0.96	0.96	0.95	0.97	0.95	0.97	0.97	0.98	0.97	0.95	0.96	0.95	0.95	0.97	0.98	0.98	0.00	0.95	0.98	0.97	1.00	0.95	0.98	0.98	0.97	0.96	0.97	0.98	0.98	0.98	0.96
0.74	0.66	0.69	0.70	0.66	0.90	0.75	0.83	0.69	0.99	0.75	0.65	0.75	0.70	0.96	0.71	0.96	0.74	0.72	0.95	0.00	0.78	0.72	1.00	0.60	0.72	0.73	0.73	0.70	0.74	0.87	0.76	0.77	0.82
0.69	0.79	0.80	0.72	0.75	0.93	0.81	0.93	0.72	0.99	0.63	0.80	0.74	0.74	0.93	0.82	0.99	0.57	0.67	0.98	0.78	0.00	0.56	1.00	0.78	0.58	0.54	0.84	0.61	0.49	0.78	0.73	0.74	0.75
0.60	0.67	0.75	0.71	0.72	0.92	0.78	0.88	0.67	0.99	0.55	0.69	0.74	0.66	0.94	0.73	0.99	0.55	0.63	0.97	0.72	0.56	0.00	0.99	0.76	0.46	0.42	0.79	0.46	0.50	0.66	0.61	0.70	0.80
0.99	0.98	0.99	1.00	1.00	0.99	0.99	1.00	0.99	1.00	0.99	0.99	0.99	0.99	1.00	0.99	1.00	0.98	0.99	1.00	1.00	1.00	0.99	0.00	0.99	0.99	0.99	0.98	0.99	0.97	0.98	0.99	0.99	1.00
0.75	0.68	0.70	0.72	0.63	0.89	0.75	0.79	0.63	1.00	0.78	0.61	0.70	0.67	0.96	0.70	0.95	0.72	0.72	0.95	0.60	0.78	0.76	0.99	0.00	0.78	0.78	0.66	0.72	0.76	0.89	0.79	0.76	0.81
0.62	0.73	0.77	0.73	0.74	0.94	0.81	0.89	0.71	0.99	0.54	0.72	0.76	0.72	0.92	0.78	0.99	0.60	0.66	0.98	0.72	0.58	0.46	0.99	0.78	0.00	0.48	0.78	0.53	0.52	0.70	0.63	0.73	0.80
0.61	0.70	0.77	0.72	0.73	0.90	0.80	0.90	0.72	1.00	0.56	0.72	0.76	0.71	0.93	0.76	0.99	0.57	0.64	0.98	0.73	0.54	0.42	0.99	0.78	0.48	0.00	0.83	0.51	0.50	0.69	0.64	0.73	0.81
0.80	0.70	0.71	0.78	0.75	0.87	0.76	0.77	0.71	0.99	0.82	0.67	0.80	0.69	0.96	0.68	0.95	0.79	0.76	0.97	0.73	0.84	0.79	0.98	0.66	0.78	0.83	0.00	0.74	0.81	0.89	0.80	0.79	0.85
0.63	0.65	0.73	0.67	0.70	0.91	0.75	0.85	0.66	0.99	0.56	0.68	0.72	0.65	0.95	0.72	0.99	0.59	0.64	0.96	0.70	0.61	0.46	0.99	0.72	0.53	0.51	0.74	0.00	0.57	0.74	0.65	0.69	0.79
0.66	0.74	0.78	0.68	0.72	0.90	0.80	0.89	0.69	1.00	0.61	0.71	0.73	0.72	0.93	0.79	0.98	0.49	0.61	0.97	0.74	0.49	0.50	0.97	0.76	0.52	0.50	0.81	0.57	0.00	0.64	0.68	0.70	0.78
0.78	0.85	0.87	0.82	0.88	0.94	0.90	0.92	0.86	0.99	0.77	0.80	0.91	0.85	0.92	0.90	0.99	0.68	0.79	0.98	0.87	0.78	0.66	0.98	0.89	0.70	0.69	0.89	0.74	0.64	0.00	0.76	0.82	0.90
0.58	0.74	0.78	0.77	0.76	0.92	0.79	0.91	0.74	0.98	0.62	0.74	0.82	0.74	0.94	0.78	0.98	0.69	0.73	0.98	0.76	0.73	0.61	0.99	0.79	0.63	0.64	0.80	0.65	0.68	0.76	0.00	0.80	0.87
0.77	0.75	0.78	0.62	0.69	0.91	0.81	0.87	0.65	0.99	0.77	0.68	0.63	0.69	0.96	0.78	0.98	0.66	0.60	0.98	0.77	0.74	0.70	0.99	0.76	0.73	0.73	0.79	0.69	0.70	0.82	0.80	0.00	0.69
0.86	0.79	0.83	0.71	0.68	0.92	0.78	0.91	0.69	0.99	0.83	0.85	0.60	0.74	0.94	0.83	0.95	0.79	0.72	0.96	0.82	0.75	0.80	1.00	0.81	0.80	0.81	0.85	0.79	0.78	0.90	0.87	0.69	0.00
0.75	0.69	0.75	0.75	0.73	0.90	0.77	0.85	0.70	0.99	0.75	0.72	0.75	0.70	0.95	0.76	0.97	0.73	0.73	0.97	0.74	0.76	0.69	0.99	0.74	0.72	0.72	0.78	0.70	0.73	0.84	0.76	0.76	0.80

Figure 11.1: Cluster Analysis of Ethnic Groups Based on the Dissimilarity Index, Toronto

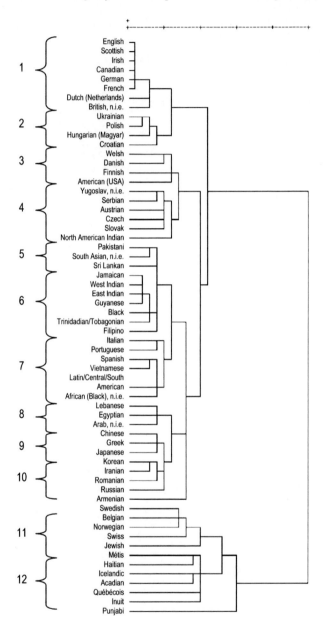

of groups rather than individual groups. This approach utilizes the same information as included in Table 11.1, but attempts to compensate for for the size of the table and the enormity of the information by making the inherent patterns more detectable to the 'inexperienced eye.' To facilitate this, we have included below a dendrogram, which is a graphical way to illustrate the same information. Created through a statistical procedure called cluster analysis, dendrograms demonstrate a gradual process of moving from one extreme scenario to another, whereas the former allows for each ethnic group to be a cluster of its own, and the latter treats all groups as belonging to one unified cluster. Between these two extremes, there exists a wide range of possibilities, each containing several clusters, in each of which one can find several groups; the groups that have the lowest values of DI – that is, the least dissimilar, or most similar groups – end up in the same cluster.

The distribution of ethnic groups, as illustrated in Figure 11.1, reveals the presence of at least 12 more-or-less meaningful clusters. The first cluster includes the groups of western European origins, namely, English, Scottish, Irish, Canadian, German, and French; as a second tier to this cluster, they are also joined by the Dutch and the British, n.i.e. (not included elsewhere). As noted in the previous paragraphs, the values of DI for these groups were extremely small, making this cluster a relatively strong and cohesive one. A second cluster includes four eastern European groups – i.e., Ukrainian, Polish, Hungarian, and Croatian – with the closest connection existing between the first two groups. The third cluster also links four groups, combining the Welsh and Americans with two groups of Nordic origins – Danish and Finnish. This is an interesting cluster, as it indicates some separation between Welsh and the rest of the British origin groups, and also a similar separation between the Nordic groups from two other Scandinavian groups – Swedish and Norwegian – which have appeared in cluster 11, alongside Belgian, Swiss, and Jewish.

Several other groups of eastern European origins have appeared in the 4th cluster, which includes Yugoslav, Serbian, Czech, Slovak, Austrian, and North American Indian. Like a couple of other clusters in this series, this one seems to be a mixed bag containing groups that do not fit well within their natural clusters. Clusters 5 and 6 seem to be representing two sets of visible minority groups with close geographical proximity; the former including groups with origins in the Indian subcontinent –

Table 11.2: Dissimilarity Index, Vancouver

	Canadian	English	French	Scottish	Irish	German	Italian	Chinese	Ukrainian	North American Indian	Dutch (Netherlands)	Polish	East Indian	Norwegian	Portuguese	Welsh	Jewish	Russian	Filipino	Métis	Swedish	Hungarian (Magyar)	American (USA)	Greek	Spanish	Jamaican	Danish
Canadian	0.00	0.16	0.26	0.18	0.21	0.17	0.45	0.57	0.23	0.55	0.26	0.33	0.57	0.32	0.48	0.49	0.65	0.43	0.47	0.63	0.40	0.37	0.59	0.52	0.54	0.71	0.36
English	0.16	0.00	0.28	0.15	0.19	0.18	0.48	0.58	0.25	0.59	0.30	0.36	0.61	0.34	0.53	0.48	0.61	0.42	0.52	0.67	0.39	0.39	0.58	0.52	0.58	0.73	0.37
French	0.26	0.28	0.00	0.28	0.28	0.30	0.50	0.60	0.31	0.56	0.36	0.37	0.63	0.40	0.55	0.55	0.62	0.46	0.52	0.64	0.45	0.41	0.62	0.55	0.55	0.71	0.44
Scottish	0.18	0.15	0.28	0.00	0.20	0.20	0.48	0.58	0.25	0.58	0.32	0.35	0.60	0.35	0.52	0.47	0.61	0.44	0.50	0.66	0.40	0.38	0.58	0.52	0.56	0.72	0.37
Irish	0.21	0.19	0.28	0.20	0.00	0.23	0.48	0.59	0.28	0.58	0.32	0.35	0.62	0.34	0.54	0.50	0.61	0.43	0.53	0.65	0.40	0.38	0.57	0.54	0.56	0.71	0.39
German	0.17	0.18	0.30	0.20	0.23	0.00	0.47	0.56	0.25	0.57	0.28	0.36	0.55	0.33	0.50	0.49	0.65	0.43	0.50	0.66	0.40	0.39	0.60	0.53	0.56	0.72	0.37
Italian	0.45	0.48	0.50	0.48	0.48	0.47	0.00	0.50	0.46	0.65	0.55	0.48	0.64	0.55	0.48	0.67	0.73	0.58	0.56	0.68	0.57	0.54	0.73	0.61	0.61	0.76	0.58
Chinese	0.57	0.58	0.60	0.58	0.59	0.56	0.50	0.00	0.55	0.65	0.66	0.58	0.62	0.67	0.48	0.73	0.57	0.57	0.42	0.76	0.66	0.58	0.73	0.59	0.54	0.75	0.65
Ukrainian	0.23	0.25	0.31	0.25	0.28	0.25	0.46	0.55	0.00	0.56	0.35	0.33	0.58	0.34	0.49	0.53	0.63	0.42	0.46	0.66	0.41	0.39	0.63	0.55	0.50	0.51	0.49
North American Indian	0.55	0.59	0.56	0.58	0.58	0.57	0.65	0.65	0.56	0.00	0.62	0.58	0.69	0.64	0.63	0.74	0.78	0.65	0.56	0.69	0.67	0.56	0.74	0.72	0.60	0.75	0.63
Dutch (Netherlands)	0.26	0.30	0.36	0.32	0.32	0.28	0.55	0.66	0.35	0.62	0.00	0.45	0.63	0.40	0.56	0.53	0.70	0.51	0.58	0.66	0.47	0.47	0.64	0.60	0.63	0.75	0.45
Polish	0.33	0.36	0.37	0.35	0.35	0.36	0.48	0.58	0.33	0.58	0.45	0.00	0.59	0.41	0.50	0.57	0.66	0.49	0.46	0.66	0.48	0.41	0.66	0.57	0.52	0.66	0.48
East Indian	0.57	0.61	0.63	0.60	0.62	0.55	0.64	0.62	0.58	0.69	0.63	0.59	0.00	0.64	0.57	0.75	0.80	0.65	0.50	0.75	0.66	0.64	0.82	0.68	0.63	0.70	0.65
Norwegian	0.32	0.34	0.40	0.35	0.34	0.33	0.55	0.67	0.34	0.64	0.40	0.41	0.64	0.00	0.57	0.51	0.72	0.50	0.58	0.63	0.45	0.44	0.64	0.60	0.62	0.75	0.42
Portuguese	0.48	0.53	0.55	0.52	0.54	0.50	0.48	0.48	0.49	0.63	0.56	0.50	0.57	0.57	0.00	0.70	0.76	0.59	0.47	0.70	0.60	0.55	0.75	0.62	0.54	0.72	0.59
Welsh	0.49	0.48	0.55	0.47	0.50	0.49	0.67	0.73	0.53	0.74	0.53	0.57	0.75	0.51	0.70	0.00	0.70	0.61	0.69	0.74	0.55	0.59	0.64	0.65	0.68	0.76	0.58
Jewish	0.65	0.61	0.62	0.61	0.61	0.65	0.73	0.57	0.63	0.78	0.70	0.66	0.80	0.72	0.76	0.70	0.00	0.61	0.68	0.85	0.72	0.65	0.73	0.59	0.70	0.77	0.73
Russian	0.43	0.42	0.46	0.44	0.43	0.43	0.58	0.57	0.42	0.65	0.51	0.49	0.65	0.50	0.59	0.61	0.61	0.00	0.54	0.72	0.55	0.48	0.67	0.61	0.59	0.70	0.53
Filipino	0.47	0.52	0.52	0.50	0.53	0.50	0.56	0.42	0.46	0.56	0.58	0.46	0.50	0.58	0.47	0.69	0.68	0.54	0.00	0.71	0.60	0.49	0.76	0.61	0.48	0.68	0.58
Métis	0.63	0.67	0.64	0.66	0.65	0.66	0.68	0.76	0.66	0.69	0.66	0.66	0.75	0.63	0.70	0.74	0.85	0.72	0.71	0.00	0.67	0.65	0.73	0.80	0.71	0.76	0.68
Swedish	0.40	0.39	0.45	0.40	0.40	0.40	0.57	0.66	0.41	0.67	0.47	0.48	0.66	0.45	0.60	0.55	0.72	0.55	0.60	0.67	0.00	0.50	0.62	0.63	0.62	0.73	0.51
Hungarian (Magyar)	0.37	0.39	0.41	0.38	0.38	0.39	0.54	0.58	0.39	0.56	0.47	0.41	0.64	0.44	0.55	0.59	0.65	0.48	0.49	0.65	0.50	0.00	0.63	0.58	0.57	0.71	0.49
American (USA)	0.59	0.58	0.62	0.58	0.57	0.60	0.73	0.73	0.63	0.74	0.64	0.66	0.82	0.64	0.75	0.64	0.73	0.67	0.76	0.73	0.62	0.63	0.00	0.67	0.73	0.81	0.63
Greek	0.52	0.52	0.55	0.52	0.54	0.53	0.61	0.59	0.55	0.72	0.60	0.57	0.68	0.60	0.62	0.65	0.59	0.61	0.61	0.80	0.63	0.58	0.67	0.00	0.64	0.76	0.62
Spanish	0.54	0.58	0.55	0.56	0.56	0.56	0.61	0.54	0.54	0.60	0.63	0.52	0.63	0.62	0.54	0.68	0.70	0.59	0.48	0.71	0.62	0.57	0.73	0.64	0.00	0.72	0.60
Jamaican	0.71	0.73	0.71	0.72	0.71	0.72	0.76	0.75	0.70	0.75	0.75	0.66	0.70	0.75	0.72	0.76	0.77	0.70	0.68	0.76	0.73	0.71	0.81	0.76	0.72	0.00	0.74
Danish	0.36	0.37	0.44	0.37	0.39	0.37	0.58	0.65	0.41	0.63	0.45	0.48	0.65	0.42	0.59	0.58	0.73	0.53	0.58	0.68	0.51	0.49	0.63	0.62	0.60	0.74	0.00
Vietnamese	0.66	0.71	0.67	0.70	0.71	0.68	0.61	0.52	0.65	0.61	0.73	0.68	0.60	0.73	0.52	0.82	0.83	0.74	0.48	0.76	0.74	0.68	0.82	0.71	0.59	0.78	0.73
British, n.i.e.	0.40	0.39	0.42	0.40	0.41	0.41	0.57	0.63	0.45	0.63	0.49	0.49	0.69	0.49	0.61	0.56	0.61	0.54	0.60	0.69	0.49	0.47	0.63	0.58	0.61	0.74	0.54
Austrian	0.47	0.45	0.53	0.45	0.47	0.45	0.62	0.66	0.49	0.68	0.51	0.52	0.71	0.51	0.65	0.53	0.67	0.56	0.64	0.70	0.52	0.56	0.61	0.65	0.65	0.75	0.52
Lebanese	0.75	0.75	0.74	0.74	0.74	0.76	0.73	0.78	0.75	0.81	0.78	0.69	0.81	0.76	0.75	0.74	0.81	0.72	0.77	0.81	0.77	0.74	0.80	0.76	0.74	0.71	0.75
Romanian	0.60	0.61	0.59	0.57	0.59	0.60	0.67	0.69	0.56	0.71	0.64	0.57	0.77	0.59	0.65	0.69	0.74	0.56	0.64	0.73	0.59	0.61	0.75	0.72	0.67	0.74	0.65
Belgian	0.83	0.82	0.83	0.82	0.81	0.82	0.88	0.86	0.82	0.83	0.80	0.85	0.89	0.81	0.88	0.83	0.87	0.82	0.86	0.83	0.83	0.80	0.81	0.87	0.85	0.85	0.81
Finnish	0.44	0.46	0.47	0.45	0.45	0.47	0.58	0.65	0.48	0.68	0.53	0.48	0.70	0.49	0.57	0.63	0.72	0.55	0.63	0.68	0.55	0.49	0.69	0.64	0.66	0.71	0.53
Swiss	0.62	0.59	0.62	0.60	0.60	0.60	0.73	0.77	0.63	0.76	0.63	0.68	0.82	0.69	0.79	0.67	0.70	0.69	0.77	0.80	0.64	0.65	0.63	0.69	0.77	0.81	0.66
Korean	0.49	0.49	0.52	0.50	0.48	0.51	0.54	0.55	0.48	0.68	0.57	0.44	0.68	0.54	0.58	0.67	0.70	0.52	0.55	0.74	0.57	0.52	0.72	0.64	0.60	0.72	0.57
Québécois	0.96	0.97	0.96	0.97	0.96	0.97	0.98	0.97	0.96	0.93	0.97	0.97	0.96	0.96	0.98	0.97	0.95	0.97	0.96	0.96	0.97	0.96	0.97	0.95	0.97	0.92	0.97
African (Black), n.i.e.	0.75	0.77	0.75	0.76	0.74	0.75	0.76	0.73	0.76	0.75	0.81	0.73	0.75	0.76	0.71	0.80	0.83	0.78	0.69	0.80	0.76	0.75	0.81	0.79	0.75	0.78	0.79
Croatian	0.54	0.57	0.58	0.57	0.58	0.56	0.44	0.54	0.55	0.68	0.62	0.57	0.66	0.61	0.56	0.69	0.74	0.59	0.57	0.71	0.62	0.59	0.80	0.61	0.62	0.76	0.62
Iranian	0.60	0.56	0.61	0.56	0.58	0.60	0.64	0.66	0.59	0.69	0.68	0.57	0.78	0.65	0.71	0.68	0.73	0.59	0.66	0.80	0.63	0.61	0.70	0.71	0.69	0.80	0.63
Japanese	0.45	0.44	0.44	0.43	0.43	0.45	0.49	0.44	0.42	0.63	0.55	0.46	0.65	0.55	0.55	0.62	0.55	0.47	0.51	0.73	0.53	0.47	0.66	0.57	0.54	0.72	0.55
Haitian	0.98	0.98	0.97	0.98	0.98	0.98	0.99	0.96	0.98	0.99	0.98	0.98	0.96	0.97	0.99	0.98	0.95	0.97	0.98	0.99	0.99	0.98	0.98	0.98	0.96	0.98	0.99
Czech	0.50	0.50	0.53	0.50	0.51	0.52	0.63	0.68	0.50	0.69	0.56	0.54	0.74	0.56	0.67	0.62	0.69	0.58	0.62	0.71	0.58	0.55	0.64	0.66	0.66	0.75	0.59
Icelandic	0.73	0.72	0.73	0.73	0.73	0.73	0.82	0.82	0.74	0.83	0.72	0.74	0.87	0.69	0.84	0.73	0.79	0.77	0.82	0.77	0.72	0.76	0.78	0.78	0.83	0.85	0.77
Pakistani	0.70	0.72	0.73	0.72	0.71	0.70	0.75	0.72	0.71	0.80	0.73	0.68	0.55	0.72	0.70	0.80	0.82	0.73	0.69	0.80	0.74	0.74	0.85	0.81	0.73	0.77	0.76
Arab, n.i.e.	0.78	0.79	0.75	0.77	0.77	0.78	0.83	0.74	0.76	0.79	0.81	0.71	0.77	0.78	0.81	0.78	0.80	0.74	0.73	0.80	0.79	0.73	0.83	0.83	0.73	0.76	0.79
Acadian	0.96	0.97	0.95	0.96	0.96	0.97	0.97	0.97	0.96	0.93	0.96	0.97	0.99	0.96	0.98	0.98	0.96	0.96	0.97	0.95	0.97	0.96	0.94	0.97	0.97	0.98	0.96
Yugoslav, n.i.e.	0.67	0.67	0.66	0.68	0.65	0.67	0.67	0.65	0.67	0.75	0.73	0.65	0.78	0.71	0.69	0.72	0.73	0.63	0.69	0.80	0.68	0.65	0.76	0.72	0.67	0.76	0.71
Sri Lankan	0.84	0.85	0.85	0.85	0.85	0.83	0.85	0.77	0.83	0.83	0.87	0.78	0.69	0.84	0.74	0.81	0.88	0.78	0.69	0.87	0.83	0.79	0.87	0.84	0.75	0.79	0.85
West Indian	0.86	0.86	0.86	0.87	0.87	0.86	0.90	0.90	0.86	0.87	0.88	0.87	0.87	0.86	0.87	0.84	0.89	0.83	0.88	0.87	0.84	0.84	0.88	0.87	0.88	0.85	0.86
Inuit	0.98	0.98	0.97	0.98	0.98	0.98	0.99	0.99	0.98	0.90	0.98	0.97	0.99	0.99	0.99	0.99	0.99	0.97	0.98	0.98	0.99	0.96	0.98	0.97	0.95	0.96	0.98
Serbian	0.75	0.74	0.71	0.72	0.71	0.75	0.74	0.73	0.70	0.78	0.79	0.67	0.85	0.74	0.74	0.80	0.76	0.64	0.73	0.86	0.74	0.70	0.83	0.78	0.71	0.77	0.76
Black	0.85	0.86	0.83	0.86	0.86	0.87	0.84	0.84	0.86	0.83	0.86	0.81	0.88	0.84	0.84	0.85	0.84	0.84	0.81	0.75	0.84	0.82	0.90	0.90	0.81	0.79	0.84
Guyanese	0.95	0.95	0.96	0.96	0.96	0.96	0.97	0.95	0.94	0.94	0.96	0.94	0.97	0.97	0.95	0.93	0.95	0.95	0.95	0.96	0.97	0.95	0.96	0.96	0.94	0.93	0.95
Slovak	0.65	0.64	0.66	0.65	0.65	0.65	0.71	0.76	0.63	0.78	0.70	0.64	0.78	0.65	0.72	0.70	0.76	0.66	0.75	0.76	0.68	0.65	0.73	0.75	0.71	0.78	0.66
Trinidadian/Tobagonian	0.86	0.87	0.83	0.86	0.85	0.87	0.90	0.89	0.87	0.87	0.87	0.84	0.88	0.87	0.87	0.91	0.92	0.89	0.84	0.91	0.85	0.86	0.90	0.91	0.89	0.83	0.88
South Asian, n.i.e.	0.72	0.74	0.75	0.73	0.74	0.71	0.78	0.71	0.72	0.80	0.76	0.72	0.47	0.75	0.67	0.79	0.82	0.71	0.63	0.83	0.75	0.73	0.87	0.77	0.72	0.71	0.77
Punjabi	0.75	0.79	0.79	0.78	0.79	0.75	0.81	0.79	0.75	0.81	0.77	0.78	0.41	0.78	0.69	0.85	0.90	0.77	0.66	0.81	0.80	0.78	0.89	0.82	0.75	0.79	0.79
Latin/Central/South American, n.i.e.	0.72	0.75	0.71	0.73	0.74	0.72	0.71	0.68	0.71	0.70	0.79	0.68	0.69	0.74	0.69	0.76	0.83	0.70	0.61	0.73	0.72	0.70	0.80	0.76	0.61	0.72	0.72
Egyptian	0.89	0.90	0.88	0.90	0.89	0.90	0.93	0.91	0.90	0.92	0.92	0.89	0.94	0.90	0.93	0.89	0.89	0.88	0.90	0.92	0.88	0.89	0.91	0.94	0.92	0.94	0.91
Armenian	0.85	0.82	0.85	0.82	0.83	0.83	0.85	0.86	0.85	0.83	0.86	0.84	0.90	0.85	0.88	0.84	0.84	0.81	0.86	0.93	0.80	0.82	0.77	0.80	0.84	0.85	0.84
Average	0.59	0.60	0.61	0.60	0.60	0.60	0.68	0.69	0.60	0.72	0.65	0.62	0.72	0.64	0.68	0.71	0.76	0.65	0.66	0.76	0.66	0.63	0.75	0.71	0.69	0.77	0.66

Source: 2001 Census Profile, CMA - CA - CT level data, Ethnic tabulations - single responses.

Vietnamese	British, n.i.e.	Austrian	Lebanese	Romanian	Belgian	Finnish	Swiss	Korean	Québécois	African (Black), n.i.e.	Croatian	Iranian	Japanese	Haitian	Czech	Icelandic	Pakistani	Arab, n.i.e.	Acadian	Yugoslav, n.i.e.	Sri Lankan	West Indian	Inuit	Serbian	Black	Guyanese	Slovak	Trinidadian/Tobagonian	South Asian, n.i.e.	Punjabi	Latin/Central/South American, n.i.e.	Egyptian	Armenian
0.66	0.40	0.47	0.75	0.60	0.83	0.44	0.62	0.49	0.96	0.75	0.54	0.60	0.45	0.98	0.50	0.73	0.70	0.78	0.96	0.67	0.84	0.86	0.98	0.75	0.85	0.95	0.65	0.86	0.72	0.75	0.72	0.89	0.85
0.71	0.39	0.45	0.75	0.61	0.82	0.46	0.59	0.49	0.97	0.77	0.57	0.56	0.44	0.98	0.50	0.72	0.72	0.79	0.97	0.67	0.85	0.86	0.98	0.74	0.86	0.95	0.64	0.87	0.74	0.79	0.75	0.90	0.82
0.67	0.42	0.53	0.74	0.59	0.83	0.47	0.62	0.52	0.96	0.75	0.58	0.61	0.44	0.97	0.53	0.73	0.73	0.75	0.95	0.66	0.85	0.86	0.97	0.71	0.83	0.96	0.66	0.83	0.75	0.79	0.71	0.88	0.85
0.70	0.40	0.45	0.74	0.57	0.82	0.45	0.60	0.50	0.97	0.76	0.57	0.56	0.43	0.98	0.50	0.73	0.72	0.77	0.96	0.68	0.85	0.87	0.98	0.72	0.86	0.96	0.65	0.86	0.73	0.78	0.73	0.90	0.82
0.71	0.41	0.47	0.74	0.59	0.81	0.45	0.60	0.48	0.96	0.74	0.58	0.58	0.43	0.98	0.51	0.73	0.71	0.77	0.96	0.65	0.85	0.87	0.98	0.71	0.86	0.96	0.65	0.85	0.74	0.79	0.74	0.89	0.83
0.68	0.41	0.45	0.76	0.60	0.82	0.47	0.60	0.51	0.97	0.75	0.56	0.60	0.45	0.98	0.52	0.73	0.70	0.78	0.97	0.67	0.83	0.86	0.98	0.75	0.87	0.96	0.65	0.87	0.71	0.75	0.72	0.90	0.83
0.61	0.57	0.62	0.73	0.67	0.88	0.58	0.73	0.54	0.98	0.76	0.44	0.64	0.49	0.99	0.63	0.82	0.75	0.83	0.97	0.67	0.85	0.90	0.99	0.74	0.84	0.97	0.71	0.90	0.78	0.81	0.71	0.93	0.85
0.52	0.63	0.66	0.78	0.69	0.86	0.65	0.77	0.55	0.97	0.73	0.54	0.66	0.44	0.96	0.68	0.82	0.72	0.74	0.97	0.65	0.77	0.90	0.99	0.73	0.84	0.95	0.76	0.89	0.71	0.79	0.68	0.91	0.86
0.50	0.45	0.49	0.75	0.56	0.82	0.48	0.63	0.48	0.96	0.76	0.55	0.59	0.42	0.98	0.50	0.74	0.71	0.76	0.96	0.67	0.83	0.86	0.98	0.70	0.86	0.94	0.63	0.87	0.72	0.75	0.71	0.90	0.85
0.61	0.63	0.68	0.81	0.71	0.83	0.68	0.76	0.68	0.93	0.75	0.68	0.69	0.63	0.99	0.69	0.83	0.80	0.79	0.93	0.75	0.83	0.87	0.90	0.78	0.83	0.94	0.78	0.87	0.80	0.81	0.70	0.92	0.83
0.73	0.49	0.51	0.78	0.64	0.80	0.53	0.63	0.57	0.97	0.81	0.62	0.68	0.55	0.98	0.56	0.72	0.73	0.81	0.96	0.73	0.87	0.88	0.98	0.79	0.86	0.96	0.70	0.87	0.76	0.77	0.79	0.92	0.86
0.68	0.49	0.52	0.69	0.57	0.85	0.48	0.68	0.44	0.97	0.73	0.57	0.57	0.46	0.98	0.54	0.74	0.68	0.71	0.97	0.65	0.78	0.87	0.97	0.67	0.81	0.94	0.64	0.84	0.72	0.78	0.68	0.89	0.84
0.60	0.69	0.71	0.81	0.77	0.89	0.70	0.82	0.68	0.96	0.75	0.66	0.78	0.65	0.96	0.74	0.87	0.55	0.77	0.99	0.78	0.69	0.87	0.99	0.85	0.88	0.97	0.78	0.88	0.47	0.41	0.69	0.94	0.90
0.73	0.49	0.51	0.76	0.59	0.81	0.49	0.69	0.54	0.96	0.76	0.61	0.65	0.55	0.97	0.56	0.69	0.72	0.78	0.96	0.71	0.84	0.86	0.99	0.74	0.84	0.97	0.65	0.87	0.75	0.78	0.74	0.90	0.85
0.52	0.61	0.65	0.75	0.65	0.88	0.57	0.79	0.58	0.98	0.71	0.56	0.71	0.55	0.99	0.67	0.84	0.70	0.81	0.98	0.69	0.74	0.87	0.99	0.74	0.84	0.95	0.72	0.87	0.67	0.69	0.69	0.93	0.88
0.82	0.56	0.53	0.74	0.69	0.83	0.63	0.67	0.67	0.97	0.80	0.69	0.68	0.62	0.98	0.62	0.73	0.80	0.78	0.98	0.72	0.81	0.84	0.99	0.80	0.85	0.93	0.70	0.91	0.79	0.85	0.76	0.89	0.84
0.83	0.61	0.67	0.81	0.74	0.87	0.72	0.70	0.70	0.95	0.83	0.74	0.73	0.55	0.95	0.69	0.79	0.82	0.80	0.96	0.73	0.88	0.89	0.99	0.76	0.84	0.95	0.76	0.92	0.82	0.90	0.83	0.89	0.84
0.74	0.54	0.56	0.72	0.56	0.82	0.55	0.69	0.52	0.97	0.78	0.59	0.59	0.47	0.97	0.58	0.77	0.73	0.74	0.96	0.63	0.78	0.83	0.97	0.64	0.84	0.95	0.66	0.89	0.71	0.77	0.70	0.88	0.81
0.48	0.60	0.64	0.77	0.64	0.86	0.63	0.77	0.55	0.96	0.69	0.57	0.66	0.51	0.98	0.62	0.82	0.69	0.73	0.97	0.69	0.69	0.88	0.98	0.73	0.81	0.95	0.75	0.84	0.63	0.66	0.61	0.90	0.86
0.76	0.69	0.70	0.81	0.73	0.83	0.68	0.80	0.74	0.96	0.80	0.71	0.80	0.73	0.99	0.71	0.77	0.80	0.80	0.95	0.80	0.87	0.87	0.98	0.86	0.75	0.96	0.76	0.91	0.83	0.81	0.73	0.92	0.93
0.74	0.49	0.52	0.77	0.59	0.83	0.55	0.64	0.57	0.97	0.76	0.62	0.63	0.53	0.99	0.58	0.72	0.74	0.79	0.97	0.68	0.83	0.84	0.99	0.74	0.84	0.97	0.68	0.85	0.75	0.80	0.72	0.88	0.80
0.68	0.47	0.56	0.74	0.61	0.80	0.49	0.65	0.52	0.96	0.75	0.59	0.61	0.47	0.98	0.55	0.76	0.74	0.73	0.96	0.65	0.79	0.84	0.96	0.70	0.82	0.95	0.65	0.86	0.73	0.78	0.70	0.89	0.82
0.82	0.63	0.61	0.80	0.75	0.81	0.69	0.63	0.72	0.97	0.81	0.80	0.70	0.66	0.98	0.64	0.78	0.85	0.83	0.94	0.76	0.87	0.88	0.98	0.83	0.90	0.96	0.73	0.90	0.87	0.89	0.80	0.91	0.77
0.71	0.58	0.65	0.76	0.72	0.87	0.64	0.69	0.64	0.95	0.79	0.61	0.71	0.57	0.98	0.66	0.78	0.81	0.83	0.97	0.72	0.84	0.87	0.97	0.78	0.90	0.96	0.75	0.91	0.77	0.82	0.76	0.94	0.80
0.59	0.61	0.65	0.74	0.67	0.85	0.66	0.77	0.60	0.97	0.75	0.62	0.69	0.54	0.96	0.66	0.83	0.73	0.73	0.97	0.67	0.75	0.88	0.95	0.71	0.81	0.94	0.71	0.89	0.72	0.75	0.61	0.92	0.84
0.78	0.74	0.75	0.71	0.74	0.85	0.71	0.81	0.72	0.92	0.78	0.76	0.80	0.72	0.98	0.75	0.85	0.77	0.76	0.98	0.76	0.79	0.85	0.96	0.77	0.79	0.93	0.78	0.83	0.71	0.79	0.72	0.94	0.85
0.73	0.54	0.52	0.75	0.65	0.81	0.53	0.66	0.57	0.97	0.79	0.62	0.63	0.55	0.99	0.59	0.77	0.76	0.79	0.96	0.71	0.85	0.86	0.98	0.76	0.84	0.95	0.66	0.88	0.77	0.79	0.72	0.91	0.84
0.00	0.73	0.77	0.85	0.79	0.90	0.73	0.86	0.73	0.96	0.78	0.65	0.83	0.69	0.98	0.77	0.89	0.78	0.81	0.98	0.77	0.74	0.88	0.97	0.83	0.86	0.97	0.83	0.92	0.74	0.72	0.64	0.97	0.95
0.73	0.00	0.55	0.74	0.67	0.81	0.57	0.61	0.59	0.96	0.74	0.66	0.63	0.50	0.98	0.58	0.71	0.75	0.79	0.95	0.70	0.86	0.86	0.96	0.74	0.85	0.97	0.66	0.87	0.79	0.85	0.74	0.88	0.79
0.77	0.55	0.00	0.76	0.65	0.76	0.61	0.61	0.56	0.96	0.76	0.62	0.65	0.55	0.98	0.56	0.71	0.75	0.76	0.97	0.68	0.87	0.85	0.97	0.74	0.84	0.96	0.66	0.87	0.77	0.81	0.73	0.90	0.79
0.85	0.74	0.76	0.00	0.72	0.89	0.75	0.77	0.74	0.98	0.80	0.73	0.78	0.73	0.99	0.75	0.83	0.80	0.83	0.98	0.75	0.84	0.83	0.92	0.74	0.81	0.99	0.76	0.88	0.85	0.87	0.75	0.93	0.88
0.79	0.67	0.65	0.72	0.00	0.83	0.64	0.77	0.61	0.98	0.80	0.64	0.64	0.63	0.98	0.64	0.80	0.77	0.74	0.96	0.69	0.84	0.82	0.97	0.67	0.79	0.97	0.69	0.84	0.78	0.86	0.76	0.86	0.82
0.90	0.81	0.76	0.89	0.83	0.00	0.83	0.73	0.83	0.95	0.86	0.87	0.82	0.82	0.95	0.85	0.85	0.90	0.84	0.97	0.87	0.89	0.88	0.98	0.83	0.89	0.95	0.83	0.87	0.90	0.92	0.86	0.85	0.83
0.73	0.57	0.61	0.75	0.64	0.83	0.00	0.71	0.60	0.97	0.77	0.61	0.65	0.57	0.98	0.61	0.75	0.75	0.78	0.98	0.71	0.85	0.83	0.99	0.75	0.83	0.94	0.67	0.84	0.75	0.83	0.77	0.88	0.86
0.86	0.61	0.61	0.77	0.77	0.73	0.71	0.00	0.72	0.97	0.80	0.78	0.67	0.67	0.99	0.66	0.73	0.81	0.80	0.94	0.78	0.88	0.89	0.97	0.80	0.88	0.99	0.73	0.88	0.82	0.89	0.80	0.89	0.78
0.73	0.59	0.56	0.74	0.61	0.83	0.60	0.72	0.00	0.97	0.75	0.57	0.56	0.48	0.98	0.57	0.74	0.71	0.69	0.96	0.66	0.79	0.87	0.97	0.65	0.82	0.92	0.67	0.84	0.78	0.82	0.70	0.89	0.80
0.96	0.96	0.96	0.98	0.98	0.95	0.97	0.97	0.97	0.00	0.96	0.98	0.99	0.95	0.91	0.94	0.96	0.95	0.98	0.86	0.96	0.93	0.97	0.94	0.98	0.96	0.95	0.95	0.99	0.97	0.96	0.98	0.97	1.00
0.78	0.74	0.76	0.80	0.80	0.86	0.77	0.80	0.75	0.96	0.00	0.73	0.80	0.77	0.97	0.78	0.83	0.75	0.78	0.98	0.75	0.81	0.89	0.98	0.82	0.86	0.95	0.76	0.78	0.80	0.81	0.74	0.95	0.90
0.65	0.66	0.62	0.73	0.64	0.87	0.61	0.78	0.57	0.98	0.73	0.00	0.66	0.55	0.97	0.68	0.82	0.71	0.80	0.97	0.64	0.84	0.86	0.98	0.72	0.84	0.97	0.70	0.88	0.74	0.79	0.69	0.95	0.86
0.83	0.63	0.65	0.78	0.64	0.82	0.65	0.67	0.56	0.99	0.80	0.66	0.00	0.58	0.98	0.65	0.77	0.80	0.72	0.97	0.70	0.86	0.89	0.99	0.70	0.87	0.96	0.67	0.90	0.82	0.91	0.75	0.84	0.71
0.69	0.50	0.55	0.73	0.63	0.82	0.57	0.67	0.48	0.95	0.77	0.55	0.58	0.00	0.97	0.54	0.76	0.73	0.74	0.95	0.64	0.82	0.85	0.98	0.63	0.82	0.97	0.66	0.90	0.73	0.83	0.70	0.88	0.81
0.98	0.98	0.98	0.99	0.98	0.95	0.98	0.99	0.98	0.91	0.97	0.97	0.98	0.97	0.00	0.97	0.97	0.96	0.97	1.00	0.97	1.00	0.96	1.00	0.98	0.98	1.00	0.96	0.97	0.98	0.94	0.95	0.97	0.99
0.77	0.58	0.56	0.75	0.64	0.85	0.61	0.66	0.57	0.94	0.78	0.68	0.65	0.54	0.97	0.00	0.74	0.79	0.76	0.96	0.69	0.85	0.87	0.98	0.71	0.81	0.93	0.66	0.86	0.77	0.83	0.73	0.88	0.78
0.89	0.71	0.71	0.83	0.80	0.85	0.75	0.73	0.74	0.96	0.83	0.82	0.77	0.76	0.97	0.74	0.00	0.85	0.84	0.95	0.82	0.85	0.88	0.99	0.85	0.80	0.96	0.76	0.88	0.88	0.89	0.88	0.93	0.80
0.78	0.75	0.75	0.80	0.77	0.90	0.75	0.81	0.71	0.95	0.75	0.71	0.80	0.73	0.96	0.79	0.85	0.00	0.80	0.98	0.80	0.79	0.84	0.98	0.83	0.87	0.98	0.77	0.85	0.65	0.64	0.73	0.93	0.92
0.81	0.79	0.76	0.83	0.74	0.84	0.78	0.80	0.69	0.98	0.78	0.80	0.72	0.74	0.97	0.76	0.84	0.80	0.00	0.98	0.79	0.83	0.93	0.97	0.76	0.84	0.94	0.78	0.84	0.75	0.81	0.80	0.90	0.89
0.98	0.95	0.97	0.98	0.96	0.97	0.98	0.94	0.96	0.86	0.98	0.97	0.97	0.95	1.00	0.96	0.95	0.98	0.98	0.00	0.97	1.00	0.94	0.92	0.95	0.93	1.00	0.96	0.99	0.98	0.97	0.96	0.96	0.98
0.77	0.70	0.68	0.75	0.69	0.87	0.71	0.78	0.66	0.96	0.75	0.64	0.70	0.64	0.97	0.69	0.82	0.80	0.79	0.97	0.00	0.86	0.89	0.99	0.65	0.85	0.96	0.71	0.89	0.78	0.85	0.73	0.92	0.85
0.74	0.86	0.87	0.84	0.84	0.89	0.85	0.88	0.79	0.93	0.81	0.84	0.86	0.82	1.00	0.85	0.85	0.79	0.83	1.00	0.86	0.00	0.89	0.98	0.88	0.83	1.00	0.86	0.86	0.75	0.74	0.81	0.88	0.87
0.88	0.86	0.85	0.83	0.82	0.88	0.83	0.89	0.87	0.97	0.89	0.86	0.89	0.85	0.96	0.87	0.88	0.84	0.93	0.94	0.89	0.89	0.00	0.93	0.85	0.82	0.91	0.86	0.90	0.87	0.89	0.87	0.92	0.91
0.97	0.96	0.97	0.92	0.97	0.98	0.99	0.97	0.97	0.94	0.98	0.98	0.99	0.98	1.00	0.98	0.99	0.98	0.97	0.92	0.99	0.98	0.93	0.00	0.99	0.92	1.00	1.00	0.92	0.99	0.99	0.98	0.98	1.00
0.83	0.74	0.74	0.74	0.67	0.83	0.75	0.80	0.65	0.98	0.82	0.72	0.70	0.63	0.98	0.71	0.85	0.83	0.76	0.95	0.65	0.88	0.85	0.99	0.00	0.87	0.97	0.70	0.89	0.88	0.93	0.71	0.89	0.83
0.86	0.85	0.84	0.81	0.79	0.89	0.83	0.88	0.82	0.96	0.86	0.84	0.87	0.82	0.98	0.81	0.80	0.87	0.84	0.93	0.85	0.83	0.82	0.92	0.87	0.00	0.95	0.84	0.89	0.84	0.90	0.81	0.90	0.92
0.97	0.97	0.96	0.99	0.97	0.95	0.94	0.99	0.92	0.95	0.95	0.97	0.96	0.97	1.00	0.93	0.96	0.98	0.94	1.00	0.96	1.00	0.91	1.00	0.97	0.95	0.00	0.93	0.97	0.97	0.98	0.98	0.94	0.96
0.83	0.66	0.66	0.76	0.69	0.83	0.67	0.73	0.67	0.95	0.76	0.70	0.67	0.66	0.96	0.66	0.76	0.77	0.78	0.96	0.71	0.86	0.86	1.00	0.70	0.84	0.93	0.00	0.86	0.80	0.86	0.76	0.88	0.81
0.92	0.87	0.87	0.88	0.84	0.87	0.84	0.88	0.84	0.99	0.78	0.88	0.90	0.90	0.97	0.86	0.88	0.85	0.84	0.99	0.89	0.86	0.90	0.92	0.89	0.89	0.97	0.86	0.00	0.87	0.88	0.91	0.95	0.94
0.74	0.79	0.77	0.85	0.78	0.90	0.75	0.82	0.78	0.97	0.80	0.74	0.82	0.73	0.98	0.77	0.88	0.65	0.75	0.98	0.78	0.75	0.87	0.99	0.88	0.84	0.97	0.80	0.87	0.00	0.53	0.78	0.93	0.91
0.72	0.85	0.81	0.87	0.86	0.92	0.83	0.89	0.82	0.96	0.81	0.79	0.91	0.83	0.94	0.83	0.89	0.64	0.81	0.97	0.85	0.74	0.89	0.99	0.93	0.90	0.98	0.86	0.88	0.53	0.00	0.75	0.96	0.93
0.64	0.74	0.73	0.75	0.76	0.86	0.77	0.80	0.70	0.98	0.74	0.69	0.75	0.70	0.95	0.73	0.88	0.73	0.80	0.96	0.73	0.81	0.87	0.98	0.71	0.81	0.98	0.76	0.91	0.78	0.75	0.00	0.94	0.88
0.97	0.88	0.90	0.93	0.86	0.85	0.88	0.89	0.89	0.97	0.95	0.95	0.84	0.88	0.97	0.88	0.93	0.93	0.90	0.96	0.92	0.88	0.92	0.98	0.89	0.90	0.94	0.88	0.95	0.93	0.96	0.94	0.00	0.85
0.95	0.79	0.79	0.88	0.82	0.83	0.86	0.78	0.80	1.00	0.90	0.86	0.71	0.81	0.99	0.78	0.80	0.92	0.89	0.98	0.85	0.87	0.91	1.00	0.83	0.92	0.96	0.81	0.94	0.91	0.93	0.88	0.85	0.00
0.76	0.66	0.67	0.79	0.72	0.85	0.68	0.75	0.67	0.96	0.80	0.70	0.72	0.64	0.98	0.69	0.81	0.78	0.80	0.96	0.75	0.84	0.87	0.97	0.78	0.85	0.96	0.75	0.88	0.78	0.82	0.77	0.91	0.84

Pakistani, South Asian, and Sri Lankan. Interestingly enough, East Indians are not closely related to this cluster, and have instead become a part of the next one that combines those from central America and the Caribbean, such as Jamaican, West Indian, Guyanese, Trinidadian and Tobagonian, and Blacks; most likely, the commonalities with West Indians has been a factor in this alignment. Filipino also is included in this cluster.

Cluster 7 has put together some southern European groups – Italian, Portuguese, Spanish – with those of Latin, Central, South American origins, as well as the residuals of the African Blacks (n.i.e.). The presence of some historical linkages alludes to the possibility of some cultural commonalities among these groups. The only mismatch here is the Vietnamese. The opposite image of this combination can be found in cluster 9, which has combined Chinese and Japanese, but also Greek. Between these two unusual combinations, there exists the 8th cluster, which includes only those of Arab origins – Lebanese, Egyptian, and other Arabs.

Cluster 12 is one with a common element, the Inuit culture, and cluster 10 seems to be another mixed bag, except perhaps for the two geographically close groups of Russian and Romanian. The two groups of Armenian and Punjabi do not fit very well in any of the existing clusters. This is an interesting phenomenon, given the supposedly similar cultural traits between Punjabis and other groups from the Indian subcontinent, and also that of Armenians and other Christian groups; history definitely needs to be consulted in search of some clues here.

The patterns observed above do not necessarily repeat themselves in a similar way in other cities. For one thing, the list of the groups present makes a difference in terms of whom each group would feel closer to. For another, the dynamics of social life is different in different cities, particularly those with distinct histories, such as the cities in eastern and western Canada. Some such differences would be immediately clear from comparing the Toronto pattern with that of, say, Vancouver, whose information is reported in Table 11.2 below and the corresponding Figure 11.2.

Figure 11.2: Cluster Analysis of Ethnic Groups Based on the Dissimilarity Index, Vancouver

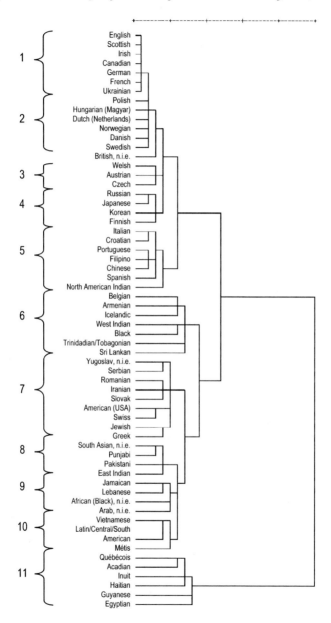

Here again, let's start with a quick review of some individual groups, using Toronto patterns as the baseline. In terms of the first two clusters observed in Toronto – one consisting of 'Canadian' and several other ethnic groups of western European origins; and the other, those of Eastern European origins – a similar situation can be found in Vancouver, with only two exceptions: Ukrainians are part of the first, rather than the second cluster; and two groups of Nordic origins – Danish and Norwegian – are also a part of the second cluster. The groups that are the most segregated from these first two clusters are Quebecois and Haitian, both predominantly French-speaking.

The composition of the clusters formed in Vancouver shows some similarities and some differences with those found in Toronto. In general, though, it seems that the boundaries between clusters are not formed along cultural and/or geographical lines, at least not to the extent as was the case in Toronto. Despite this, there exist a few clusters with clear-cut boundaries, the first one of which is the so-called western European cluster. The groups included in this cluster are: English, Scottish, Irish, Canadian, German, and French, but also, interestingly, Ukrainian. This cluster is located very close to a second cluster, which can be called northern-eastern European cluster, including most Nordic groups – Swedish, Danish, Norwegian – but also the Dutch, and two eastern European groups of Polish and Hungarian.

There are a few other clusters that contain more or less similar groups in terms of their cultural and/or geographical features. Examples are: Japanese and Koreans in cluster 4; southern European groups – Italian, Portuguese, and Spanish – in cluster 5; Yugoslav, Serbian, Romanian, and Slovak in cluster 7; and Jamaican and African on the one hand, and Lebanese and other Arabs, on the other, both in cluster 9. The only other cluster with a relatively strong and visible common trait is cluster 8, which includes Punjabi, Pakistani, East Indian, and South Asian groups.

Two groups with unusually high degrees of segregation in Vancouver are Quebecois and Haitian, both predominantly French-speaking. The question that might arise here is: to what extent can this be attributed to the minority status of these groups in a predominantly English-speaking environment, and to what extent is it a possible tendency among the Francophone populations? Table 11.3 contains the DI values for all ethnic groups in Montreal. It reveals noticeably higher degrees of segregation between the French and almost all the rest of the ethnic groups. The only exception is their segregation from 'Canadian' (0.19) and, to a lesser extent, Irish (0.34).

It should be noted, however, that the composition of 'Canadian' in Montreal is different from the other two cities, as in the former, it consists mostly of the Francophones. In a sense, this would mean that there is a close cultural similarity between the French and the 'Canadian' groups in Montreal, hence, an extremely low level of segregation between the two. Here again, most of the smaller groups have higher values of DI; but, even the larger groups have unusually higher values as well, at least compared to Vancouver and Toronto. This latter fact has contributed to a distinct situation in Montreal with a noticeably high degree of segregation amongst all groups, an interesting contrast to add to the mix examined so far.

In addition to the high values of DI, the clusters of groups found in Montreal (Figure 11.3) are also at odds with those found elsewhere. The strongest cluster – that is, one with extremely low DI values among its constituting groups – is what can be called the French block, consisting of the French, Canadian, and Quebecois. At later stages, these groups are joined by Belgian – a partly Francophone group – Irish, and the Aboriginals. The closest cluster to this, in terms of strength, is what is perhaps the equivalent to the western European cluster in other cities; the ethnic groups that appear in this cluster include: English, Scottish, German, but also three eastern European groups of Polish, Hungarian, and Ukrainian.

Arabs – including Lebanese, Egyptian, and others – make a relatively distinct cluster of their own. A similar thing can be seen for Sri Lankan, Pakistani, Punjabi, and South Asian, all from the Indian subcontinent. Another surprising thing about the spatial trends in Montreal is that all the Scandinavian groups – Danish, Finnish, Norwegian, and Swedish – have ended up in the same cluster. A similarly distinct cluster can be found for Métis, Icelandic, and Inuit, who are of similar cultural background, and also for Jamaican, West Indian, and Trinidadian/Tobagonian, although the latter are joined by Filipino and Croatian.

Table 11.3: Dissimilarity Index, Montréal

	Canadian	English	French	Scottish	Irish	German	Italian	Chinese	Ukrainian	North American Indian	Dutch (Netherlands)	Polish	East Indian	Norwegian	Portuguese	Welsh	Jewish	Russian	Filipino	Métis	Swedish	Hungarian (Magyar)	American (USA)	Greek	Spanish	Jamaican	Danish
Canadian	0.00	0.60	0.19	0.54	0.35	0.55	0.62	0.67	0.69	0.50	0.78	0.65	0.79	0.96	0.56	0.98	0.91	0.81	0.85	0.86	0.95	0.73	0.82	0.76	0.59	0.84	0.95
English	0.60	0.00	0.55	0.35	0.43	0.42	0.67	0.59	0.56	0.69	0.60	0.47	0.64	0.85	0.72	0.89	0.77	0.66	0.70	0.93	0.88	0.55	0.73	0.69	0.62	0.66	0.85
French	0.19	0.55	0.00	0.50	0.34	0.51	0.60	0.62	0.64	0.53	0.76	0.60	0.75	0.96	0.53	0.97	0.87	0.75	0.82	0.88	0.94	0.69	0.81	0.72	0.54	0.82	0.95
Scottish	0.54	0.35	0.50	0.00	0.41	0.46	0.68	0.65	0.59	0.68	0.61	0.53	0.70	0.86	0.70	0.88	0.81	0.72	0.75	0.91	0.87	0.60	0.75	0.74	0.65	0.73	0.83
Irish	0.35	0.43	0.34	0.41	0.00	0.45	0.61	0.61	0.60	0.55	0.68	0.53	0.71	0.92	0.61	0.94	0.85	0.74	0.77	0.87	0.90	0.64	0.76	0.72	0.58	0.76	0.91
German	0.55	0.42	0.51	0.46	0.45	0.00	0.68	0.62	0.61	0.67	0.63	0.51	0.65	0.88	0.68	0.91	0.78	0.69	0.74	0.91	0.89	0.56	0.75	0.65	0.62	0.71	0.88
Italian	0.62	0.67	0.60	0.68	0.61	0.68	0.00	0.62	0.64	0.75	0.82	0.63	0.73	0.95	0.55	0.96	0.88	0.77	0.83	0.91	0.94	0.77	0.85	0.73	0.55	0.81	0.96
Chinese	0.67	0.59	0.62	0.65	0.61	0.62	0.62	0.00	0.63	0.75	0.76	0.55	0.55	0.95	0.65	0.94	0.77	0.64	0.68	0.93	0.93	0.69	0.79	0.64	0.56	0.66	0.94
Ukrainian	0.69	0.56	0.64	0.59	0.60	0.61	0.64	0.63	0.00	0.74	0.72	0.49	0.68	0.93	0.69	0.92	0.81	0.65	0.75	0.92	0.93	0.65	0.78	0.73	0.48	0.52	0.57
North American Indian	0.50	0.69	0.53	0.68	0.55	0.67	0.75	0.75	0.74	0.00	0.79	0.74	0.84	0.96	0.71	0.98	0.93	0.83	0.87	0.86	0.95	0.79	0.82	0.85	0.71	0.84	0.96
Dutch (Netherlands)	0.78	0.60	0.76	0.61	0.68	0.63	0.82	0.76	0.72	0.79	0.00	0.67	0.80	0.88	0.82	0.88	0.83	0.76	0.83	0.90	0.83	0.66	0.77	0.83	0.79	0.79	0.82
Polish	0.65	0.47	0.60	0.53	0.53	0.51	0.63	0.55	0.49	0.74	0.67	0.00	0.59	0.91	0.68	0.91	0.73	0.62	0.69	0.91	0.91	0.55	0.75	0.64	0.58	0.62	0.91
East Indian	0.79	0.64	0.75	0.70	0.71	0.65	0.73	0.55	0.68	0.84	0.80	0.59	0.00	0.92	0.75	0.95	0.76	0.68	0.64	0.97	0.94	0.71	0.83	0.55	0.66	0.58	0.93
Norwegian	0.96	0.85	0.96	0.86	0.92	0.88	0.95	0.95	0.93	0.96	0.88	0.91	0.92	0.00	0.96	0.84	0.94	0.91	0.94	0.94	0.87	0.91	0.93	0.96	0.95	0.94	0.81
Portuguese	0.56	0.72	0.53	0.70	0.61	0.68	0.55	0.65	0.69	0.71	0.82	0.68	0.75	0.96	0.00	0.98	0.89	0.79	0.86	0.89	0.93	0.75	0.85	0.72	0.55	0.84	0.97
Welsh	0.98	0.89	0.97	0.88	0.94	0.91	0.96	0.94	0.92	0.98	0.88	0.91	0.95	0.84	0.98	0.00	0.95	0.96	0.96	1.00	0.85	0.91	0.93	0.93	0.97	0.96	0.80
Jewish	0.91	0.77	0.87	0.81	0.85	0.78	0.88	0.77	0.81	0.93	0.83	0.73	0.76	0.94	0.89	0.95	0.00	0.66	0.74	0.99	0.93	0.69	0.82	0.77	0.83	0.74	0.91
Russian	0.81	0.66	0.75	0.72	0.74	0.69	0.77	0.64	0.65	0.83	0.76	0.62	0.68	0.91	0.79	0.96	0.66	0.00	0.66	0.94	0.88	0.65	0.79	0.77	0.67	0.64	0.91
Filipino	0.85	0.70	0.82	0.75	0.77	0.74	0.83	0.68	0.75	0.87	0.83	0.69	0.64	0.94	0.86	0.96	0.74	0.66	0.00	0.98	0.94	0.75	0.85	0.79	0.77	0.60	0.92
Métis	0.86	0.93	0.88	0.91	0.87	0.91	0.91	0.93	0.92	0.86	0.90	0.91	0.97	0.94	0.89	1.00	0.99	0.94	0.98	0.00	0.95	0.91	0.88	0.96	0.90	0.96	0.95
Swedish	0.95	0.88	0.94	0.87	0.90	0.89	0.94	0.93	0.93	0.95	0.83	0.91	0.94	0.87	0.93	0.85	0.93	0.88	0.94	0.95	0.00	0.85	0.92	0.96	0.93	0.95	0.85
Hungarian (Magyar)	0.73	0.55	0.69	0.60	0.64	0.56	0.77	0.69	0.65	0.79	0.66	0.55	0.71	0.91	0.75	0.91	0.69	0.65	0.75	0.91	0.85	0.00	0.72	0.69	0.70	0.71	0.90
American (USA)	0.82	0.73	0.81	0.75	0.76	0.75	0.85	0.79	0.78	0.82	0.77	0.75	0.83	0.93	0.85	0.93	0.82	0.79	0.85	0.88	0.92	0.72	0.00	0.86	0.80	0.85	0.91
Greek	0.76	0.69	0.72	0.74	0.72	0.65	0.73	0.64	0.73	0.85	0.83	0.64	0.55	0.96	0.72	0.93	0.77	0.77	0.79	0.96	0.96	0.69	0.86	0.00	0.70	0.76	0.95
Spanish	0.59	0.62	0.54	0.65	0.58	0.62	0.55	0.56	0.63	0.71	0.79	0.58	0.66	0.95	0.55	0.97	0.83	0.67	0.77	0.90	0.93	0.70	0.80	0.70	0.00	0.72	0.95
Jamaican	0.84	0.66	0.82	0.73	0.76	0.71	0.81	0.66	0.70	0.84	0.79	0.62	0.58	0.94	0.84	0.96	0.74	0.64	0.60	0.96	0.95	0.71	0.85	0.76	0.72	0.00	0.91
Danish	0.95	0.85	0.95	0.83	0.91	0.88	0.96	0.94	0.91	0.96	0.82	0.91	0.93	0.81	0.97	0.80	0.91	0.91	0.92	0.95	0.85	0.90	0.91	0.95	0.95	0.91	0.00
Vietnamese	0.71	0.72	0.66	0.75	0.71	0.71	0.67	0.52	0.71	0.77	0.84	0.65	0.62	0.95	0.63	0.96	0.79	0.71	0.69	0.94	0.93	0.73	0.86	0.70	0.55	0.72	0.96
British, n.i.e.	0.90	0.67	0.88	0.70	0.79	0.73	0.89	0.79	0.79	0.89	0.69	0.73	0.80	0.84	0.92	0.79	0.78	0.80	0.86	0.92	0.84	0.72	0.81	0.86	0.88	0.81	0.79
Austrian	0.89	0.74	0.88	0.76	0.83	0.77	0.89	0.83	0.80	0.90	0.81	0.78	0.79	0.86	0.90	0.94	0.83	0.81	0.86	0.93	0.93	0.75	0.86	0.82	0.87	0.83	0.85
Lebanese	0.72	0.69	0.68	0.73	0.70	0.63	0.65	0.64	0.73	0.81	0.81	0.65	0.68	0.94	0.70	0.95	0.77	0.70	0.79	0.95	0.93	0.66	0.83	0.57	0.64	0.77	0.95
Romanian	0.70	0.60	0.64	0.65	0.64	0.61	0.70	0.61	0.62	0.76	0.75	0.54	0.66	0.93	0.70	0.95	0.71	0.54	0.69	0.93	0.88	0.61	0.80	0.69	0.62	0.66	0.93
Belgian	0.51	0.67	0.49	0.66	0.57	0.65	0.76	0.75	0.70	0.65	0.75	0.69	0.81	0.95	0.70	0.97	0.88	0.78	0.85	0.89	0.93	0.71	0.81	0.79	0.69	0.84	0.93
Finnish	0.98	0.85	0.96	0.87	0.93	0.88	0.96	0.91	0.90	0.96	0.83	0.90	0.91	0.85	0.96	0.84	0.88	0.86	0.92	1.00	0.89	0.87	0.89	0.95	0.96	0.87	0.80
Swiss	0.81	0.73	0.78	0.74	0.78	0.73	0.87	0.84	0.79	0.82	0.73	0.77	0.84	0.89	0.83	0.93	0.84	0.81	0.89	0.93	0.87	0.74	0.80	0.85	0.81	0.89	0.90
Korean	0.88	0.72	0.85	0.73	0.80	0.76	0.85	0.72	0.78	0.88	0.74	0.74	0.79	0.92	0.89	0.89	0.81	0.73	0.85	0.96	0.88	0.75	0.78	0.87	0.82	0.79	0.88
Québécois	0.36	0.66	0.34	0.62	0.47	0.62	0.65	0.67	0.71	0.56	0.82	0.67	0.79	0.97	0.58	0.98	0.90	0.79	0.87	0.85	0.94	0.76	0.80	0.79	0.60	0.85	0.97
African (Black), n.i.e.	0.77	0.73	0.72	0.76	0.73	0.74	0.73	0.62	0.72	0.79	0.84	0.68	0.64	0.96	0.75	0.98	0.86	0.66	0.71	0.93	0.91	0.76	0.85	0.75	0.62	0.65	0.96
Croatian	0.88	0.75	0.87	0.78	0.81	0.75	0.76	0.79	0.79	0.86	0.82	0.75	0.78	0.91	0.81	0.92	0.84	0.80	0.83	0.91	0.92	0.75	0.86	0.75	0.78	0.76	0.91
Iranian	0.83	0.64	0.78	0.69	0.72	0.65	0.81	0.68	0.71	0.87	0.72	0.60	0.70	0.87	0.82	0.91	0.74	0.60	0.74	0.94	0.86	0.62	0.75	0.76	0.73	0.72	0.86
Japanese	0.92	0.75	0.90	0.79	0.85	0.80	0.88	0.79	0.86	0.92	0.78	0.80	0.82	0.86	0.89	0.91	0.87	0.83	0.88	0.94	0.84	0.78	0.83	0.87	0.85	0.87	0.86
Haitian	0.64	0.77	0.62	0.77	0.69	0.73	0.48	0.69	0.75	0.75	0.87	0.71	0.79	0.97	0.58	0.99	0.92	0.80	0.85	0.88	0.95	0.80	0.86	0.78	0.54	0.83	0.98
Czech	0.90	0.74	0.88	0.75	0.82	0.80	0.91	0.79	0.80	0.88	0.76	0.75	0.80	0.89	0.91	0.94	0.78	0.78	0.83	0.92	0.89	0.74	0.82	0.85	0.86	0.77	0.89
Icelandic	1.00	0.98	0.99	0.98	0.99	0.98	0.99	0.98	0.98	0.98	0.97	0.97	0.98	0.95	0.98	0.97	0.97	0.98	0.99	0.97	0.99	0.97	0.96	0.99	0.98	0.98	0.98
Pakistani	0.90	0.80	0.88	0.85	0.85	0.79	0.87	0.70	0.82	0.91	0.87	0.76	0.57	0.96	0.86	0.98	0.89	0.83	0.80	0.95	0.97	0.80	0.90	0.64	0.82	0.76	0.96
Arab, n.i.e.	0.73	0.67	0.68	0.70	0.69	0.64	0.63	0.57	0.68	0.78	0.79	0.61	0.63	0.93	0.65	0.95	0.80	0.66	0.75	0.92	0.92	0.66	0.83	0.64	0.58	0.72	0.95
Acadian	0.72	0.78	0.71	0.78	0.74	0.77	0.82	0.81	0.77	0.72	0.82	0.78	0.87	0.95	0.74	0.95	0.93	0.86	0.92	0.91	0.93	0.82	0.85	0.88	0.75	0.86	0.93
Yugoslav, n.i.e.	0.91	0.76	0.89	0.81	0.85	0.81	0.84	0.78	0.79	0.89	0.82	0.76	0.80	0.93	0.87	0.95	0.83	0.74	0.84	0.91	0.91	0.76	0.85	0.84	0.80	0.79	0.91
Sri Lankan	0.92	0.86	0.90	0.89	0.89	0.84	0.89	0.77	0.85	0.91	0.90	0.80	0.61	0.97	0.89	0.99	0.81	0.76	0.64	0.94	0.97	0.83	0.91	0.72	0.80	0.73	0.98
West Indian	0.91	0.73	0.90	0.78	0.82	0.78	0.86	0.73	0.78	0.88	0.82	0.71	0.65	0.91	0.88	0.94	0.82	0.72	0.70	0.97	0.93	0.80	0.86	0.81	0.80	0.57	0.87
Inuit	0.96	0.95	0.96	0.94	0.96	0.96	0.97	0.94	0.96	0.94	0.95	0.96	0.96	0.93	0.95	1.00	0.99	0.95	0.93	0.93	0.95	0.96	0.95	0.98	0.96	0.92	0.88
Serbian	0.95	0.82	0.93	0.85	0.89	0.87	0.92	0.83	0.87	0.94	0.86	0.85	0.82	0.87	0.92	0.91	0.85	0.82	0.83	0.93	0.89	0.81	0.87	0.90	0.86	0.80	0.84
Black	0.76	0.75	0.74	0.77	0.75	0.75	0.63	0.69	0.73	0.78	0.81	0.69	0.69	0.97	0.70	0.97	0.87	0.73	0.78	0.90	0.93	0.79	0.81	0.78	0.63	0.70	0.94
Guyanese	0.95	0.86	0.95	0.91	0.91	0.89	0.92	0.85	0.88	0.94	0.90	0.84	0.80	0.95	0.93	0.97	0.92	0.88	0.82	0.97	0.95	0.87	0.93	0.87	0.89	0.71	0.90
Slovak	0.91	0.75	0.90	0.78	0.84	0.77	0.88	0.82	0.82	0.91	0.79	0.75	0.80	0.88	0.88	0.91	0.84	0.85	0.86	0.96	0.95	0.76	0.85	0.81	0.86	0.80	0.91
Trinidadian/Tobagonian	0.91	0.73	0.90	0.78	0.82	0.78	0.87	0.75	0.79	0.89	0.80	0.73	0.68	0.90	0.87	0.92	0.84	0.77	0.73	0.93	0.90	0.78	0.84	0.82	0.83	0.58	0.89
South Asian, n.i.e.	0.94	0.84	0.92	0.90	0.91	0.85	0.89	0.77	0.86	0.91	0.91	0.82	0.63	0.96	0.88	0.98	0.86	0.82	0.69	0.95	0.95	0.88	0.92	0.73	0.80	0.72	0.96
Punjabi	0.97	0.90	0.96	0.94	0.95	0.91	0.92	0.88	0.92	0.96	0.98	0.88	0.64	0.91	0.92	0.97	0.89	0.92	0.88	0.98	1.00	0.92	0.97	0.75	0.89	0.78	0.97
Latin/Central/South American, n.i.e.	0.77	0.80	0.75	0.81	0.77	0.79	0.65	0.69	0.74	0.78	0.87	0.73	0.78	0.99	0.66	0.98	0.90	0.77	0.84	0.87	0.95	0.80	0.85	0.82	0.60	0.80	0.97
Egyptian	0.77	0.64	0.72	0.68	0.70	0.60	0.75	0.65	0.71	0.84	0.71	0.62	0.62	0.89	0.77	0.92	0.79	0.73	0.77	0.95	0.94	0.67	0.79	0.63	0.72	0.73	0.89
Armenian	0.85	0.77	0.81	0.80	0.81	0.71	0.83	0.76	0.80	0.90	0.85	0.73	0.74	0.96	0.81	0.96	0.82	0.81	0.84	0.97	0.96	0.73	0.89	0.49	0.78	0.82	0.96
Average	0.78	0.72	0.75	0.74	0.74	0.72	0.79	0.73	0.76	0.82	0.80	0.71	0.75	0.92	0.79	0.94	0.84	0.77	0.81	0.93	0.92	0.75	0.84	0.78	0.74	0.77	0.90

Source: 2001 Census Profile, CMA - CA - CT level data, Ethnic tabulations - single responses.

Vietnamese	British, n.i.e.	Austrian	Lebanese	Romanian	Belgian	Finnish	Swiss	Korean	Québécois	African (Black), n.i.e.	Croatian	Iranian	Japanese	Haitian	Czech	Icelandic	Pakistani	Arab, n.i.e.	Acadian	Yugoslav, n.i.e.	Sri Lankan	West Indian	Inuit	Serbian	Black	Guyanese	Slovak	Trinidadian/Tobagonian	South Asian, n.i.e.	Punjabi	Latin/Central/South American, n.i.e.	Egyptian	Armenian
0.71	0.90	0.89	0.72	0.70	0.51	0.98	0.81	0.88	0.36	0.77	0.88	0.83	0.92	0.64	0.90	1.00	0.90	0.73	0.72	0.91	0.92	0.91	0.96	0.95	0.76	0.95	0.91	0.91	0.94	0.97	0.77	0.77	0.85
0.72	0.67	0.74	0.69	0.60	0.67	0.85	0.73	0.72	0.66	0.73	0.75	0.64	0.75	0.77	0.74	0.98	0.80	0.67	0.78	0.76	0.86	0.73	0.95	0.82	0.75	0.86	0.75	0.73	0.84	0.90	0.80	0.64	0.77
0.66	0.88	0.88	0.68	0.64	0.49	0.96	0.78	0.85	0.34	0.72	0.87	0.78	0.90	0.62	0.88	0.99	0.88	0.68	0.71	0.89	0.90	0.90	0.96	0.93	0.74	0.95	0.90	0.90	0.92	0.96	0.75	0.72	0.81
0.75	0.70	0.76	0.73	0.65	0.66	0.87	0.74	0.73	0.62	0.76	0.78	0.69	0.79	0.77	0.75	0.98	0.85	0.70	0.78	0.81	0.89	0.78	0.94	0.85	0.77	0.91	0.78	0.78	0.90	0.94	0.81	0.68	0.80
0.71	0.79	0.83	0.70	0.64	0.57	0.93	0.78	0.80	0.47	0.73	0.81	0.72	0.85	0.69	0.82	0.99	0.85	0.69	0.74	0.85	0.89	0.82	0.96	0.89	0.75	0.91	0.84	0.82	0.91	0.95	0.77	0.70	0.81
0.71	0.73	0.77	0.63	0.61	0.65	0.88	0.73	0.76	0.62	0.74	0.75	0.65	0.80	0.73	0.80	0.98	0.79	0.64	0.77	0.81	0.84	0.78	0.96	0.87	0.75	0.89	0.77	0.78	0.85	0.91	0.79	0.60	0.71
0.67	0.89	0.89	0.65	0.70	0.76	0.96	0.87	0.85	0.65	0.73	0.76	0.81	0.88	0.48	0.91	0.99	0.87	0.63	0.82	0.84	0.89	0.86	0.97	0.92	0.63	0.92	0.88	0.87	0.89	0.92	0.65	0.75	0.83
0.52	0.79	0.83	0.64	0.61	0.75	0.91	0.84	0.72	0.67	0.62	0.79	0.68	0.79	0.69	0.79	0.98	0.70	0.57	0.81	0.78	0.77	0.73	0.94	0.83	0.69	0.85	0.82	0.75	0.77	0.88	0.69	0.65	0.76
0.50	0.79	0.80	0.73	0.62	0.70	0.90	0.79	0.78	0.71	0.72	0.79	0.71	0.86	0.75	0.80	0.98	0.82	0.68	0.77	0.79	0.85	0.78	0.96	0.87	0.73	0.88	0.82	0.79	0.86	0.92	0.74	0.71	0.80
0.77	0.89	0.90	0.81	0.76	0.65	0.96	0.82	0.88	0.56	0.79	0.86	0.87	0.92	0.75	0.88	0.98	0.91	0.78	0.72	0.89	0.91	0.88	0.94	0.94	0.78	0.94	0.91	0.89	0.91	0.96	0.78	0.84	0.90
0.84	0.69	0.81	0.81	0.75	0.75	0.83	0.73	0.74	0.82	0.84	0.82	0.72	0.78	0.87	0.76	0.97	0.87	0.79	0.82	0.82	0.90	0.82	0.95	0.86	0.81	0.90	0.79	0.80	0.91	0.98	0.87	0.71	0.85
0.65	0.73	0.78	0.65	0.54	0.69	0.90	0.77	0.74	0.67	0.68	0.75	0.60	0.80	0.71	0.75	0.97	0.76	0.61	0.78	0.76	0.80	0.71	0.96	0.85	0.69	0.84	0.75	0.73	0.82	0.88	0.73	0.62	0.73
0.62	0.80	0.79	0.68	0.66	0.81	0.91	0.84	0.79	0.79	0.64	0.78	0.70	0.82	0.79	0.80	0.98	0.57	0.63	0.87	0.80	0.61	0.65	0.96	0.82	0.69	0.80	0.80	0.68	0.63	0.64	0.78	0.62	0.74
0.95	0.84	0.86	0.94	0.93	0.95	0.85	0.89	0.92	0.97	0.96	0.91	0.87	0.86	0.97	0.89	0.95	0.96	0.93	0.95	0.93	0.97	0.91	0.93	0.87	0.97	0.95	0.88	0.90	0.96	0.91	0.99	0.89	0.96
0.63	0.92	0.90	0.70	0.70	0.70	0.96	0.83	0.89	0.58	0.75	0.81	0.82	0.89	0.58	0.91	0.98	0.86	0.65	0.74	0.87	0.89	0.88	0.95	0.92	0.70	0.93	0.88	0.87	0.88	0.92	0.66	0.77	0.81
0.96	0.79	0.94	0.95	0.95	0.97	0.84	0.93	0.89	0.98	0.98	0.92	0.91	0.91	0.99	0.94	0.97	0.98	0.95	0.95	0.95	0.99	0.94	1.00	0.91	0.97	0.97	0.91	0.92	0.98	0.97	0.98	0.92	0.96
0.79	0.78	0.83	0.77	0.71	0.88	0.88	0.84	0.81	0.90	0.86	0.84	0.74	0.87	0.92	0.78	0.97	0.89	0.80	0.93	0.83	0.81	0.82	0.99	0.85	0.87	0.92	0.84	0.84	0.86	0.89	0.90	0.79	0.82
0.71	0.80	0.81	0.70	0.54	0.78	0.86	0.81	0.73	0.79	0.66	0.80	0.60	0.83	0.80	0.78	0.98	0.83	0.66	0.86	0.74	0.76	0.72	0.95	0.82	0.73	0.88	0.85	0.77	0.82	0.92	0.77	0.73	0.81
0.69	0.86	0.86	0.79	0.69	0.85	0.92	0.89	0.85	0.87	0.71	0.83	0.74	0.88	0.85	0.83	0.99	0.80	0.75	0.92	0.84	0.64	0.70	0.93	0.83	0.78	0.82	0.86	0.73	0.69	0.88	0.84	0.77	0.84
0.94	0.92	0.93	0.95	0.93	0.89	1.00	0.93	0.96	0.85	0.93	0.91	0.94	0.94	0.88	0.92	0.97	0.95	0.92	0.91	0.91	0.94	0.97	0.93	0.93	0.90	0.97	0.96	0.93	0.95	0.98	0.87	0.95	0.97
0.93	0.84	0.93	0.93	0.88	0.93	0.89	0.87	0.88	0.94	0.91	0.92	0.86	0.84	0.95	0.89	0.99	0.97	0.92	0.93	0.91	0.97	0.93	0.95	0.89	0.93	0.95	0.95	0.90	0.95	1.00	0.95	0.94	0.96
0.73	0.72	0.75	0.66	0.61	0.71	0.87	0.74	0.75	0.76	0.76	0.75	0.62	0.78	0.80	0.74	0.97	0.80	0.66	0.82	0.76	0.83	0.80	0.96	0.81	0.79	0.87	0.76	0.78	0.88	0.92	0.80	0.67	0.73
0.86	0.81	0.86	0.83	0.80	0.81	0.89	0.80	0.78	0.80	0.85	0.86	0.75	0.83	0.86	0.82	0.96	0.90	0.83	0.85	0.85	0.91	0.86	0.95	0.87	0.81	0.93	0.85	0.84	0.92	0.97	0.85	0.79	0.89
0.70	0.86	0.82	0.57	0.69	0.79	0.95	0.85	0.87	0.79	0.75	0.75	0.76	0.87	0.78	0.85	0.99	0.64	0.64	0.88	0.84	0.72	0.81	0.98	0.90	0.78	0.87	0.81	0.82	0.73	0.75	0.82	0.63	0.49
0.55	0.88	0.87	0.64	0.62	0.69	0.96	0.81	0.82	0.60	0.62	0.78	0.73	0.85	0.54	0.86	0.98	0.82	0.58	0.75	0.80	0.80	0.80	0.96	0.86	0.63	0.89	0.86	0.83	0.80	0.89	0.60	0.72	0.78
0.72	0.81	0.83	0.77	0.66	0.84	0.87	0.89	0.79	0.85	0.65	0.76	0.72	0.87	0.83	0.77	0.98	0.76	0.72	0.86	0.79	0.73	0.57	0.92	0.80	0.70	0.71	0.80	0.58	0.72	0.78	0.80	0.73	0.82
0.96	0.79	0.85	0.95	0.93	0.93	0.80	0.90	0.88	0.97	0.96	0.91	0.86	0.86	0.98	0.89	0.98	0.96	0.95	0.93	0.91	0.98	0.87	0.88	0.84	0.94	0.90	0.91	0.89	0.96	0.97	0.97	0.89	0.96
0.00	0.88	0.86	0.68	0.65	0.76	0.98	0.83	0.86	0.68	0.59	0.81	0.78	0.91	0.64	0.85	0.99	0.77	0.61	0.81	0.82	0.69	0.79	0.97	0.86	0.69	0.84	0.86	0.81	0.76	0.89	0.62	0.75	0.82
0.88	0.00	0.82	0.83	0.81	0.88	0.80	0.80	0.76	0.92	0.89	0.81	0.76	0.78	0.93	0.77	0.99	0.88	0.84	0.88	0.81	0.91	0.82	0.94	0.86	0.89	0.87	0.77	0.81	0.88	0.95	0.90	0.77	0.89
0.86	0.82	0.00	0.87	0.80	0.88	0.86	0.79	0.79	0.91	0.88	0.83	0.78	0.83	0.92	0.77	0.98	0.86	0.85	0.91	0.85	0.90	0.78	0.97	0.86	0.90	0.91	0.80	0.85	0.87	0.88	0.92	0.80	0.87
0.68	0.83	0.87	0.00	0.61	0.75	0.94	0.83	0.84	0.73	0.71	0.74	0.72	0.82	0.69	0.80	0.99	0.77	0.53	0.85	0.82	0.77	0.82	0.98	0.87	0.71	0.90	0.79	0.85	0.84	0.88	0.76	0.54	0.55
0.65	0.81	0.80	0.61	0.00	0.74	0.91	0.81	0.78	0.70	0.67	0.79	0.65	0.83	0.72	0.76	0.98	0.79	0.59	0.77	0.79	0.81	0.72	0.94	0.83	0.69	0.87	0.85	0.78	0.83	0.92	0.73	0.66	0.73
0.76	0.88	0.88	0.75	0.74	0.00	0.97	0.75	0.86	0.57	0.80	0.85	0.79	0.88	0.75	0.84	0.99	0.88	0.75	0.73	0.89	0.91	0.89	0.96	0.93	0.80	0.92	0.89	0.90	0.93	0.97	0.81	0.75	0.83
0.98	0.80	0.86	0.94	0.91	0.97	0.00	0.94	0.86	0.97	0.91	0.87	0.87	0.82	0.98	0.88	0.98	0.95	0.92	0.98	0.92	0.99	0.89	0.96	0.87	0.95	0.95	0.88	0.94	0.96	0.97	0.97	0.90	0.93
0.83	0.80	0.79	0.83	0.81	0.75	0.94	0.00	0.85	0.82	0.87	0.83	0.79	0.85	0.86	0.78	0.99	0.90	0.81	0.80	0.87	0.95	0.89	0.96	0.88	0.86	0.94	0.82	0.90	0.91	0.98	0.87	0.81	0.87
0.86	0.76	0.79	0.84	0.78	0.86	0.86	0.85	0.00	0.87	0.84	0.86	0.61	0.79	0.90	0.80	0.93	0.88	0.78	0.87	0.82	0.93	0.78	0.94	0.85	0.87	0.90	0.85	0.79	0.93	0.96	0.88	0.78	0.88
0.68	0.92	0.91	0.73	0.70	0.57	0.97	0.82	0.87	0.00	0.73	0.87	0.84	0.91	0.65	0.91	0.99	0.89	0.71	0.72	0.91	0.89	0.91	0.95	0.94	0.74	0.96	0.91	0.91	0.92	0.96	0.72	0.80	0.86
0.59	0.89	0.88	0.71	0.67	0.80	0.91	0.87	0.84	0.73	0.00	0.82	0.80	0.86	0.69	0.87	0.98	0.73	0.65	0.82	0.82	0.74	0.76	0.96	0.83	0.67	0.84	0.88	0.76	0.75	0.86	0.68	0.78	0.80
0.81	0.81	0.83	0.74	0.79	0.85	0.87	0.83	0.86	0.87	0.82	0.00	0.75	0.83	0.81	0.85	0.98	0.86	0.77	0.88	0.84	0.87	0.79	0.98	0.83	0.77	0.84	0.81	0.76	0.85	0.84	0.82	0.73	0.75
0.78	0.76	0.78	0.72	0.65	0.79	0.87	0.79	0.61	0.84	0.80	0.75	0.00	0.77	0.85	0.75	0.94	0.81	0.65	0.84	0.75	0.86	0.74	0.97	0.79	0.78	0.84	0.81	0.78	0.87	0.88	0.84	0.67	0.79
0.91	0.78	0.83	0.82	0.83	0.88	0.82	0.85	0.79	0.91	0.86	0.83	0.77	0.00	0.93	0.79	0.93	0.89	0.80	0.88	0.89	0.93	0.88	0.96	0.85	0.88	0.93	0.82	0.89	0.91	0.94	0.90	0.80	0.87
0.64	0.93	0.92	0.69	0.72	0.75	0.98	0.86	0.90	0.65	0.69	0.81	0.85	0.93	0.00	0.91	0.99	0.84	0.64	0.82	0.87	0.83	0.89	0.97	0.91	0.52	0.93	0.90	0.90	0.85	0.93	0.59	0.80	0.84
0.85	0.77	0.77	0.80	0.76	0.84	0.88	0.78	0.80	0.91	0.87	0.85	0.75	0.79	0.91	0.00	0.99	0.85	0.82	0.88	0.79	0.88	0.85	0.96	0.87	0.86	0.87	0.80	0.81	0.88	0.93	0.90	0.80	0.87
0.99	0.99	0.98	0.99	0.98	0.99	0.98	0.99	0.93	0.99	0.98	0.98	0.94	0.93	0.99	0.99	0.00	0.98	0.99	0.98	0.97	0.98	0.99	0.97	0.94	1.00	0.99	0.97	0.96	0.99	0.99	0.99	0.98	0.99
0.77	0.88	0.86	0.77	0.79	0.88	0.95	0.90	0.88	0.89	0.73	0.86	0.81	0.89	0.84	0.85	0.98	0.00	0.74	0.91	0.87	0.65	0.77	0.97	0.90	0.80	0.84	0.87	0.79	0.66	0.66	0.84	0.74	0.78
0.61	0.84	0.85	0.53	0.59	0.75	0.92	0.81	0.78	0.71	0.65	0.77	0.65	0.80	0.64	0.82	0.99	0.74	0.00	0.82	0.80	0.76	0.79	0.98	0.82	0.68	0.86	0.82	0.80	0.79	0.87	0.66	0.60	0.69
0.81	0.88	0.91	0.85	0.77	0.73	0.98	0.80	0.87	0.72	0.82	0.88	0.84	0.88	0.82	0.88	0.98	0.91	0.82	0.00	0.92	0.92	0.89	0.93	0.92	0.84	0.95	0.90	0.90	0.94	0.96	0.85	0.83	0.89
0.82	0.81	0.85	0.82	0.79	0.89	0.92	0.87	0.82	0.91	0.82	0.84	0.75	0.89	0.87	0.79	0.97	0.87	0.80	0.92	0.00	0.87	0.87	0.94	0.84	0.83	0.85	0.85	0.84	0.89	0.90	0.87	0.82	0.89
0.69	0.91	0.90	0.77	0.81	0.91	0.99	0.95	0.93	0.89	0.74	0.87	0.86	0.93	0.83	0.88	0.98	0.65	0.76	0.92	0.87	0.00	0.79	0.95	0.87	0.79	0.88	0.86	0.82	0.59	0.69	0.78	0.82	0.86
0.79	0.82	0.78	0.82	0.72	0.89	0.89	0.89	0.78	0.91	0.76	0.79	0.74	0.88	0.89	0.85	0.99	0.77	0.79	0.89	0.87	0.79	0.00	0.94	0.80	0.74	0.78	0.83	0.63	0.75	0.79	0.85	0.77	0.83
0.97	0.94	0.97	0.98	0.94	0.96	0.96	0.96	0.94	0.95	0.96	0.98	0.97	0.96	0.97	0.96	0.97	0.97	0.98	0.93	0.94	0.95	0.94	0.00	0.94	0.96	0.97	0.97	0.94	0.95	0.98	0.97	0.96	0.99
0.86	0.86	0.86	0.87	0.83	0.93	0.87	0.88	0.85	0.94	0.83	0.83	0.79	0.85	0.91	0.87	0.94	0.90	0.82	0.92	0.84	0.87	0.80	0.94	0.00	0.89	0.83	0.87	0.75	0.91	0.94	0.89	0.88	0.92
0.69	0.89	0.90	0.71	0.69	0.80	0.95	0.86	0.87	0.74	0.67	0.77	0.78	0.88	0.52	0.86	1.00	0.80	0.68	0.84	0.83	0.79	0.74	0.96	0.89	0.00	0.84	0.87	0.78	0.79	0.87	0.62	0.76	0.82
0.84	0.87	0.91	0.90	0.87	0.92	0.95	0.94	0.90	0.96	0.84	0.84	0.84	0.93	0.93	0.87	0.99	0.84	0.86	0.95	0.85	0.88	0.78	0.97	0.83	0.84	0.00	0.84	0.72	0.84	0.88	0.90	0.85	0.91
0.86	0.77	0.80	0.79	0.85	0.89	0.88	0.82	0.85	0.91	0.88	0.81	0.81	0.82	0.90	0.80	0.97	0.87	0.82	0.90	0.85	0.86	0.83	0.97	0.87	0.87	0.84	0.00	0.83	0.91	0.90	0.89	0.72	0.82
0.81	0.81	0.85	0.85	0.78	0.90	0.94	0.90	0.79	0.91	0.76	0.76	0.78	0.89	0.90	0.81	0.96	0.79	0.80	0.90	0.84	0.82	0.63	0.94	0.75	0.78	0.72	0.83	0.00	0.72	0.79	0.86	0.77	0.87
0.76	0.88	0.87	0.84	0.83	0.93	0.96	0.91	0.93	0.92	0.75	0.85	0.87	0.91	0.85	0.88	0.99	0.66	0.79	0.94	0.89	0.59	0.75	0.95	0.91	0.79	0.84	0.91	0.72	0.00	0.76	0.81	0.84	0.85
0.89	0.95	0.88	0.88	0.92	0.97	0.97	0.98	0.96	0.96	0.86	0.84	0.88	0.94	0.93	0.93	0.99	0.66	0.87	0.96	0.90	0.69	0.79	0.98	0.94	0.87	0.88	0.90	0.79	0.76	0.00	0.90	0.87	0.88
0.62	0.90	0.92	0.76	0.73	0.81	0.97	0.87	0.88	0.72	0.68	0.82	0.84	0.90	0.59	0.90	0.99	0.84	0.66	0.85	0.87	0.78	0.85	0.97	0.89	0.62	0.90	0.89	0.86	0.81	0.90	0.00	0.83	0.86
0.75	0.77	0.80	0.54	0.66	0.75	0.90	0.81	0.78	0.80	0.78	0.73	0.67	0.80	0.80	0.80	0.98	0.74	0.60	0.83	0.82	0.82	0.77	0.96	0.88	0.76	0.85	0.72	0.77	0.84	0.87	0.83	0.00	0.61
0.82	0.89	0.87	0.55	0.73	0.83	0.93	0.87	0.88	0.86	0.80	0.75	0.79	0.87	0.84	0.87	0.99	0.78	0.69	0.89	0.89	0.86	0.83	0.99	0.92	0.82	0.91	0.82	0.87	0.85	0.88	0.86	0.61	0.00
0.77	0.83	0.85	0.77	0.74	0.80	0.91	0.84	0.83	0.79	0.78	0.82	0.77	0.86	0.80	0.84	0.98	0.83	0.75	0.85	0.85	0.85	0.82	0.96	0.87	0.79	0.89	0.85	0.82	0.85	0.90	0.82	0.77	0.82

Figure 11.3: Cluster Analysis of Ethnic Groups Based on the Dissimilarity Index, Montreal

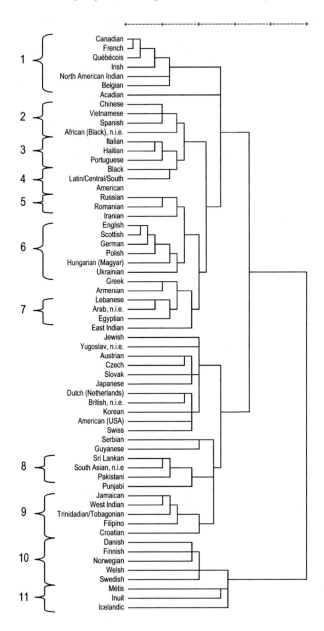

There remain many groups in Montreal that do not fit well within any of the above clusters. This includes East Indian, Yugoslav, Croatian, Korean, American, Swiss, Jewish, Acadian, Serbian, Czech, Slovak, Japanese, and Guyanese. This is despite the fact that many of these groups are large enough to be able to have a substantial presence in many neighbourhoods; hence, their high degrees of segregation cannot be attributed to their small populations. An examination of the reasons behind Montreal's distinct features is certainly beyond the scope of the present study.

Spatial Segregation and Trust

How are these spatial traits of ethnic groups related to whether or not they trust others? As mentioned earlier, the presence of a spatially-segregated environment has the potential to limit the frequency as well as the depth of social interactions. In such an environment, the interactions among people would remain limited to people of similar ethnic origin, at least as far as the neighbourhood is concerned. In a relatively homogeneous neighbourhood, therefore, the stereotypes about other ethnic groups remain unchallenged and strong, and can act as barriers to the development of unbiased attitudes towards people who belong to a different group.

If the above argument holds true, we should expect to see a negative relationship between the degree of spatial segregation of ethnic/cultural groups and the amount of trust they have in others. Figure 11.4 shows that this is exactly the case. The graph reports the average values of the Dissimilarity Index for each group and the proportion of each group who have reported having trust in others. Like the previous graphs, each dot on this graph represents a certain ethnic group in a particular city. The declining trend line in the graph indicates that the groups that have reported the highest levels of trust are those who are least segregated, and with the increase in the degree of spatial segregation their levels of trust drop.

The graph shows the countrywide trend, but similar trends were also found for each particular city, with a couple of exceptions. Of course, one major limitation in trying to zoom in on cities, as in other chapters, is that in the smaller urban centers only a few ethnic groups have a population sizeable enough to yield reliable results. Moreover, the roster of such groups varies from one city to another, because some groups have a more visible presence in one city and not in others, making the comparisons more difficult.

Figure 11.4: The relationship between ethnic groups' residential segregation (DI) and trust, 8 larger CMAs combined

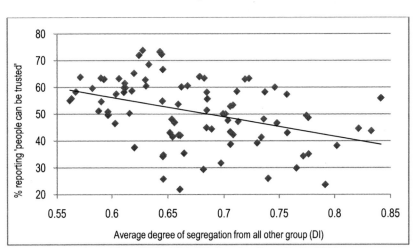

Source: Trust: EDS, 2002; DI: Canadian Census Profile, 2001

The presence of the negative relationship between the degree of segregation and trust allows us to explain a problem we noted in chapter seven, that is, the so-called 'Montreal anomaly.' The reader would remember that in chapter seven we found a strong and positive association between the degree of ethnic diversity and the level of trust in cities, a uniquely Canadian phenomenon. However, we also found that Montreal was an exception to this general trend, as the city is among the most ethnically-diverse cities in Canada, yet has a surprisingly low level of trust. In this respect, Montreal was different not only from cities elsewhere in Canada, but also from the rest of the cities in the province of

Quebec. Why, unlike the rest of Canadian cities inside and outside Quebec, has the ethnic diversity in Montreal not resulted in a higher level of trust among the city's residents?

We are still far from a position to provide a definite answer to this question, but bringing the discussion on the impact of residential segregation into the picture seems to provide a hint for at least some of reasons behind the 'Montreal anomaly.' Comparing Montreal with the two other major cities of Toronto and Vancouver in Figure 11.5 shows that, despite its high level of ethnic diversity, Montreal has a much higher degree of spatial segregation of ethnic groups than the other two. The ethnic groups in Montreal are concentrated in the lower right corner of Figure 11.5, marking the area of high-segregation and low-trust. This is in contrast to ethnic groups in Vancouver, concentrated in the top left corner, that is, the area of low-segregation and high-trust. The situation in Toronto, represented by the line in the middle, is following the general pattern seen in Montreal and Vancouver, although the strength of relationship between trust and segregation is much weaker than in the other two cities.

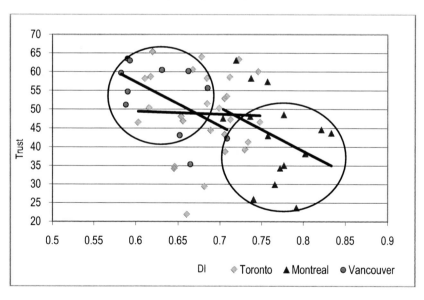

Figure 11.5: The relationship between ethnic groups' residential segregation (DI) and trust, Toronto, Montreal, Vancouver

Why should segregation matter for the relationship between diversity and trust? If ethnic diversity has a positive impact on trust, as was seen for Canadian cities, it does so through facilitating interactions among people of different ethnic backgrounds. Now, if diversity at the city level is combined with segregation at neighbourhood level, this would mean that the potentially positive impacts of exposure to diversity are suppressed by lack of diversity in neighbourhoods.

Conclusion

The findings discussed in this chapter reinforce some of patterns we noticed earlier, and shed light on an exception to those patterns. In a nutshell, the examination of the relationship between trust and the residential segregation of ethnic groups showed that those who are more segregated trust others less. This was compatible with our discussion in chapter seven, in which a positive impact between diversity and trust was found in cities, and also with our discussion in chapter eight, where we noticed a stronger tendency to trust among those who had a more heterogeneous ancestral background. The case of Montreal also showed that diversity in a broad environment may not automatically translate itself into an interaction among those with different backgrounds, ethnic or cultural.

Of course, this discussion should not be taken as suggesting that the exposure to diversity occurs only in neighbourhoods. Sure enough, such exposure can also occur in places like schools and work sites. However, the transitory nature of many contacts in those environments, compared to the longer lasting connections in neighbourhoods, can make them less influential in shaping one's attitudes towards different others. Another environment in which long-lasting and deep-seated ties have a favourable ground to grow is within religious communities. In the next chapter, we examine that aspect of diversity and its interaction with trust.

12 Trust and Religion

In the previous five chapters, several different aspects of diversity were examined in relation to trust. Here we take on a few other aspects, all of which are related to religion: belonging to and affiliation with a particular faith, strength of religious belief, and participation in religious functions. The inclusion of religion in a discussion of trust is very important, given the active role that faith and faith-based organizations play in today's world. Indeed, the rising influence of religion in shaping the political landscape of many countries over the past couple of decades makes it all the more important to see the consequences of religion on trust as an indicator of broader concepts of social capital and social cohesion.

Figure 12.1 shows the level of trust reported by Canadians of different religious backgrounds. The significant differences in the trust levels shown in the figure, along with the patterns of those differences, justify the necessity of including religion in a discussion of trust. Several points in Figure 12.1 are noteworthy. First, the difference between the groups with the highest and the lowest levels of trust is about 35 percentage points, too significant a distance to be disregarded; indeed, the score reported for the latter is half of what is reported for the former. Second, among the religious groups located at both low and high ends of the scale, one can find groups that are mostly immigrants and those that include mostly the native-born population. This feature by itself implies the complexity of the phenomenon, as it clearly indicates that the particular levels of trust reported by each group cannot be easily explained by attributing them to the local or foreign roots of those groups. Third, the two groups with the lowest levels of trust are Jehovah's Witnesses and Muslims, the former a domestic faith, and the latter consisting mostly of immigrants. So, this means that among religious groups with immigrant roots, Muslims are the least trusting religious community. This is particularly important, given that Muslims have been at the centre of the heated post-9/11 debates in the western world about diversity, ethnic minorities, national identities, security issues, and so on. A big part of these debates is a concern about the integration of second-generation Muslim immigrants in immigrant-receiving countries, and against this background, it is of utmost importance to identify the factors associated with this low level of trust among Muslims.

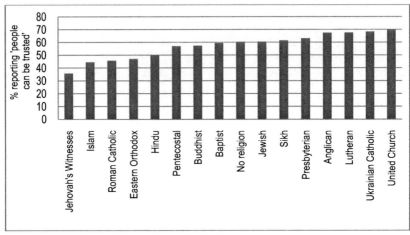

Figure 12.1: *Trust by religious background, Canada, 2003*

Source: *GSS17 (master files)*

An explanation of the above phenomenon requires an examination of factors in several different areas. One such area can be the nature of the teachings of various faiths. As one of the main functions of a religious faith is to provide the believers with a perspective on life, including both the natural world and the social, it is certain that the content of such perspectives can have far-reaching implications for the ways in which others are viewed. Another area can be the nature of the religious communities to which people belong. Durkheim, the 19th century sociologist, was among the first to distinguish between the belief and community aspects of religions, and to show the sometimes conflicting implications of the two. As far as trust is concerned, it is possible to imagine a religion that offers teachings which are not particularly conducive to trust, but which has communities capable of generating trust, and vice-versa. A third area can be the nature of encounters between a particular religious community and the larger society. Faced by the animosity of the surrounding population, for instance, members of a particular faith community can develop distrustful views towards others, contrary to the teachings of their faith. A comprehensive explanation of the trust patterns shown earlier requires a simultaneous look into all these factors.

Obviously, such a comprehensive examination of the correlates of trust in different religious communities is beyond the scope of this book. Before such a task, however, it is important to know the extent to which the above patterns are real and not proxies for other correlates of religion. Also, it is equally important to examine the interplay of trust and a couple of other aspects of religion. If nothing else, these two tasks will illuminate and further refine the research question that should be addressed in future research. Towards that goal we devote the remainder of this chapter to looking at a wide range of variables influencing trust, in order to develop a better understanding of how trustful views are developed in general, and the role that religion-related factors play in particular.

We will first discuss the conceptual framework guiding this part of the study. Then, we will discuss the results of a logistic regression model, in which we have tested the hypotheses resulting from our conceptual framework. In running this statistical model, we have utilized the data from cycle 17 of the Canadian General Social Survey.

Predictors of Trust: a Conceptual Framework

The existing literature on the determinants of social trust has pointed to many relevant variables. To make the discussion of these variables easier for the reader to follow, we have broken them down into three broad groups: 1) the standard demographic variables, such as age, sex, income, education, rural-urban residency, and immigration status; 2) variables related to communal and civic engagement, such as involvement in volunteer activities, social networks, following the news, and marital status. While the variable marital status is normally considered as one of the demographic variables, given that it speaks to the nature of social relationships within family, we found it more useful to discuss it alongside other variables in this group; and, 3) religion-related variables, which are the theme of interest in this chapter. Below, we will discuss the ways in which each of these variables is expected to influence one's tendency to trust or distrust. In addition to shedding more light on the nature of the religion-related variables, this analysis also allows for the development of a more general picture of the predictors of trust.

Demographic variables

Income
As mentioned earlier, trusting unknown others involves a certain degree of risk (Hardin, 2002, Wuthnow, 1998). When the subject of trust is someone with whom we have a long-term relationship we have some indications of the way in which they may behave towards us and this lowers the amount of uncertainty and risk involved in the interaction. In a sense, the presence of a certain degree of predictability makes it easier for the truster to calculate the associated risks. The presence of a long-term relationship also implies the presence of either a shared cultural outlook or similar personalities, both influential forces in determining the norms, including the mutual obligations, governing the relationship. In the absence of all these commonalities, trusting the unknown other becomes an unpredictable and risky business. Whether one would step into this uncertain territory would then depend, to a large extent, on their capacity to absorb those associated risks. The more the resources at one's disposal, the higher his or her capacity to tolerate such risks! Against this background, the literature suggests that a higher income can result in more trust.

Age
Putnam's (2000) study made it clear that the so-called social capitalists in the United States are most likely to be found among the older folks, or what he called the 'long civic generation' in reference to those born in the immediate post-WWII period. He also showed that many indicators of social capital have experienced a consistent decline with each subse-

quent generation. This can be taken as suggesting that age may have a positive association with trust.

There exists yet another possible reason for the increased civic engagement of older citizens, which is their higher degree of financial security and lower degree of vulnerability. This is somewhat related to a better income, but can also be related to the different kind of needs that come with age. Maslow (1943) pointed to the hierarchy of human needs, moving from the most basic ones – food, clothes, and shelter – to higher level ones such as social and economic security, the development of a sense of belonging, and eventually the need for self-actualization. An older age typically comes with a relative satisfaction of the basic needs and the establishment of a more secure and meaningful life. This status is conducive to more social engagement, and that is likely to result in more trust. In our model here, we have controlled for all these variables, to see whether or not age as a demographic factor has any other effect when the financial security and civic engagement correlates are controlled for.

Education
Of all the variables associated with social capital in general, and trust in particular, education has been most often cited as an important one with a positive impact. Those who are better educated are perceived to be better-informed about their social surroundings, possess better communication skills, enjoy higher incomes and belong to larger and more diverse social networks. These in turn weaken the influence of stereotypes in their views on others, and help them maintain a more trustful view towards people whom they do not know personally. We, therefore, expect to see the possession of more education leading to more trust in others.

Sex
The existing studies are split on the effect of gender on the strength of the tendency to trust. A few studies have suggested that women tend to trust others more than men do, and have cited the different modes of socialization – e.g., the stronger emphasis in women's upbringing on being communicative and understanding – as a possible factor. Others have shown a higher level of trust among men, and refer to their higher level of associational engagement – which is believed to result in more

trust – as the possible explanation. Women's lower average income and education have also been mentioned as relevant for the understanding of trust tendencies. The simultaneous inclusion of all these variables in our model will allows us to disentangle all these correlates, and isolate the impact of gender on trust.

Immigrant Status

As mentioned earlier, being in a position to trust others presupposes the presence of a shared history, and that is a direct product of shared time spent together by both the truster and the trusted. This shared time gives them the ability to predict each other's moves, and a capability to read each other's feelings. When such a shared personal history is not present, people can rely on a shared culture to make similar predictions. Immigrants experience a deficit in both these fronts, due to a limited history of interactions with the host population as well as inherent differences with the host culture. It is, therefore, not unrealistic to expect immigrants to express a lower level of trust in others compared to the native-born.

Shouldn't we, then, expect a similar drop in the trust levels of the native-born? Not necessarily, and that is for two main reasons. First, the 'others' who are the subjects of the distrust are not the same for immigrants and the native-born; for immigrants, it is a big population, the majority, with whom they have a short personal history and a small cultural overlap; for the native-born, it is the same large population, but only a small part of it are immigrants. Even if the native-born show distrust in this small minority, that does not necessarily lead to a distrust of the whole population. Secondly, there exist certain difficulties – with consequences for trust – that are experienced only by immigrants and not by the native-born. Many immigrants, for instance, have to start their new life in the host society with jobs and living conditions lower than what they think they deserve; many suffer from the lack of recognition of their credentials and pre-migration work experiences; some may suffer from discrimination in the job market. All these have a great potential to create a sense of alienation among immigrants towards the rest of society, resulting in distrust towards the mainstream population.

Rural-urban residency
As Wuthnow (1998) argues, life in small-scale communities creates more face-to-face contacts among the residents. The presence of a cultural homogeneity in smaller communities, in addition to the fewer but stronger number of social ties among the residents, make those communities favourable grounds for the development of trustful relationships. In contrast, the anonymity and fast pace associated with life in larger urban centers, along with the diversity and shallowness of social ties there, make it much more difficult for urban residents to develop long-lasting relationships among themselves. Pressures of time and money, which Putnam considered as the main culprits in the erosion of social capital in the U.S., have their strongest presence in larger urban centres.

Communal engagement variables

Volunteering
Putnam (2000) argues that volunteerism and trust are correlated, but the direction of the causal relationship is not well known. On the one hand, volunteerism presupposes a trust in others, as the potential volunteer must be willing to work with and for strangers; on the other, engagement in volunteer activities puts volunteers in touch with other good-intentioned volunteers, reinforcing trusting attitudes in the process. A third possibility is that both trust and volunteerism are products of a less cynical personality or a cultural trait.

In either case, it is important to see whether or not trust and volunteerism empirically correlate with each other. This, obviously, would not settle the question on the direction of causality, but it will show the possible inter-connectedness of these two very important aspects of social capital, especially when the effects of many other related variables are controlled for.

Involvement in civic activities
Putnam (2000) treats involvement in civic activities as a major factor in fostering social capital in general, and trust in particular. Given the volunteer nature of most, if not all, communal and civic activities, the nature of their impact on trust should be fairly similar to that of volunteering,

which was discussed earlier. But it is also reasonable to expect that different types of civic activities may have different types of influences on trust. Given this, several different types of engagement in civic life have been included in the model in order to examine the size and nature of the impacts of each.

Social networks

The possession and maintenance of a large social network is a difficult job. It requires a high degree of sensitivity and a great deal of interpersonal skills. The larger and denser the social network, the more complex the web of relationships, the greater the chances for conflicts to emerge among the members. Social networks also involve the investment of resources such as time, money, and emotion. But, in return for such investments, social networks offer a great deal; they can make available to members a sense of belonging, emotional support, vital information and, in many cases, even financial assistance.

Social networks consist primarily of informal ties among people and, hence, get very close to their personal and private lives. This can potentially make the members vulnerable and therefore requires some sort of informal assurance that the exposition of personal lives will not be taken advantage of. In such circumstances, a trust-based relationship is the only cement that holds the members together. Trust, in this sense, both presupposes the presence of social networks and is also reinforced by them. A positive impact of social networks on trust is therefore hypothesized here.

Marital status

Marital status, as mentioned earlier, is normally considered among the demographic variables. But, because the different types of marital statuses signify different types of social settings, and due to the fact that 'familial social capital' has been shown to have major implications for other aspects of people's lives, it would be appropriate to discuss this variable under the rubric of communal factors.

There are ample reasons to believe that marital status would interact with trust. Being married, and particularly having children, indicates the presence of a lasting relationship between at least two and possibly more people, which works fairly similarly to how a social network functions.

Being divorced, on the other hand, signifies the break-down of a trust-based relationship; the failure of the union tends to have spill-over effects in other areas of life, nurturing a suspicious attitude towards others. This effect is all the stronger in the period immediately after divorce, or in the separation phase. Being single can be taken to imply that the singles have not yet found someone in whom they can have complete trust. Common-law relationships can also be seen along the same line, except that it is perhaps one step further away from singlehood and towards married status. If our understanding of the ways in which different marital statuses interact with trust holds water, we should expect to see the highest level of trust among the married and lowest among the divorced and separated, with single and common-law in between.

Media
Putnam (2000) argued that the media played an extremely significant role in the decline of American social capital. According to him, besides things such as the pressures of time and money, suburban sprawl, and generational change, the privatization and individualization of entertainment through TV has been a main culprit in pushing civic engagement trends down. He noted that, compared to those who watch TV for entertainment, Americans who watch it for news have better social capital records, although still lower than those who get their news from newspapers.

While some question the significance that Putnam attaches to TV as one of the factors behind the civic decline in the U.S., there are others who think that some of the other factors that Putnam listed are by-products of the rising hours of TV watching. Honore (2004), for instance, blames TV – as the number one consumer of Americans' leisure time – for adding to the pressure of time because of its fast pace and the fact that it "does not allow us the time to pause or reflect" (p: 225). According to him, while taking care of life's leisure, TV takes away from life's pleasure: "heavy viewers spend less time on things that really make life pleasurable – cooking, chatting with family, exercising, making love, socializing, doing volunteer work" (p: 225). As a result of the so-called 'TV Turn-off Movement' he claims more people turn away from the TV watching habit, and towards other, more constructive, ways of spending their leisure time, such as reading.

Along similar lines, we also expect to see a lower level of trust to be associated with more time watching TV here in Canada. Given that TV provides the most passive type of entertainment, we also hypothesize that as we move towards more active media such as radio and the print press, there will be a corresponding increase in the level of trust.

Religion-related variables

Religious affiliation
Faith communities, Putnam (2000) claimed, "are arguably the single most important repository of social capital in America" (p: 66). Religious individuals have stronger philanthropic beliefs and feelings and are more widely connected to others, both factors which have the potential to foster more trust in others. But to view religion as a factor that only binds people and fosters trust would involve a huge oversight of the fact that religion has also been one of the most divisive forces throughout history. Indeed, religion has tended to move in two seemingly opposite directions: creating strong ties and high trust within the community of believers, and weakening ties and stirring strong distrust in others beyond the community.

When it comes to religion, however, society is too complex to be accounted for by a simple dichotomy of believers versus non-believers (Wuthnow, 2003). After all, there exist many different faiths and, as a consequence, many different types of believers. For that reason, to capture the essence of religion's impact on trust, it is important to go beyond that simple dichotomy and compare not only the religious with the irreligious, but also those who belong to various faith communities with each other. This will act as a proxy for the nature of the beliefs system promoted by each particular faith.

Importance of religion
Being associated with a particular faith does not say much about the strength of the association. By itself, affiliation with a religion does not show the extent to which one's life is influenced by his or her religious beliefs. As the literature on religiosity has long shown, people may use religion as a source of identity, but sometimes also as a concrete source

of direction and guidance. In the former case, religion acts mostly for the purpose of creating distinctions with others; in the latter case, it penetrates one's everyday life at a much deeper level. Given this, it would be important to see the effect of the importance given to religion on one's tendency to trust.

Religious attendance
Each religion creates a community of believers, one way or another. The level of one's engagement with this community, and the nature of those engagements, has strong consequences for how one functions in other areas of life, including the extent to which one trusts others. Attending religious functions can have the same type of impact on trust as other volunteer works and civic engagements do. It puts people in touch with those of different socioeconomic and cultural backgrounds who share the same faith. This can result in a higher level of trust in the population at large.

Who Trusts? The Results

Using the above conceptual framework, we have employed the Canadian General Social Survey (cycle 17) to examine the simultaneous effects of the three groups of variables. This has been done through a multivariate statistical technique called *logistic regression*. A detailed description of this technique is included in Appendix 12.1. Here, suffice it to say that the values reported below are the magnitudes of Exp(B), which are the estimations of the contribution of each variable to the odds of trusting others. Values smaller than 1 indicate that the variable has a suppressing effect on trust, and values greater than 1 have a reinforcing effect.

The results of the logistic regression analysis are reported in Table 12.1. The variables included in the model are arranged under the three headings proposed above, and will be discussed accordingly. The first group of factors affecting an individual's tendency to trust consists of demographic variables.

Table 12.1: Factors influencing one's tendency to trust

Demographic Factors		
Variable	Exp(B)	Sig.
Sex (being Male)	1.057	0.09
Income (in $ 10,000s)	**0.998**	0.00 *
Age	**1.003**	0.05 *
Place of residence (Rural)	**1.106**	0.01 *
Immigrant Status (being Immigrant)	1.043	0.37
Education (reference category: Incomplete High School)		
University Degree	**2.087**	0.00 *
College Degree	**1.348**	0.00 *
Some University	**1.416**	0.00 *
High School Diploma	**1.193**	0.00 *
Communal Engagement Factors		
Variable	Exp(B)	Sig.
Marital Status (reference category: Married)		
Commonlaw	**0.660**	0.00 *
Widowed	**0.819**	0.00 *
Separated	**0.685**	0.00 *
Divorced	**0.745**	0.00 *
Single	**0.661**	0.00 *
Social networks		
How many close friends do you have	**1.050**	0.00 *
How many other friends	**1.011**	0.00 *
Volunteering		
Past year: did unpaid work for a voluntary organization	**1.217**	0.00 *
Average hours/month volunteering	0.995	0.40
Civic Engagement		
Member sports recreation organization	**1.173**	0.00 *
Member service club/fraternal organization	**0.854**	0.01 *
Member union	1.032	0.46
Member political party/group	1.022	0.78
Member cultural/education/hobby organization	1.081	0.11
Member school group/neighbourhood association	1.029	0.57
Member of any other type of organization	1.057	0.41
Number of groups were a member/participant in past 12 months	1.025	0.13
Frequency of participations in group activities and meetings	1.001	0.89
Ties to local community		
Sense of belonging to local community	**1.238**	0.00 *

A neighbour has done a favour for the respondent (within the previous month)	**1.253**	0.00	*
The respondent has done a favour for a neighbour (within the previous month)	0.943	0.14	
The number of people known in the neighbourhood	1.022	0.26	
Length of time in dwelling	1.013	0.09	
Media			
The frequency in which one follows news and current affairs	**0.951**	0.01	*
The main media used for news: NEWSPAPER	**1.137**	0.00	*
The main media used for news: TELEVISION	**0.758**	0.00	*
The main media used for news: RADIO	**1.072**	0.03	*
The main media used for news: INTERNET	**1.130**	0.00	*
Hours spent watching TV	**0.996**	0.05	*
What media do you use: MAGAZINE	0.980	0.63	

Religion-related Factors			
Variable	Exp(B)	Sig.	
Religious affiliation (reference category: No Religion)			
Roman Catholic	**0.572**	0.00	*
Baptist	**0.735**	0.00	*
Eastern Orthodox	**0.583**	0.00	*
Muslim	**0.388**	0.00	*
Hindu	**0.526**	0.00	*
Pentecostal	**0.707**	0.01	*
Jehovah's Witnesses	**0.279**	0.00	*
Sikh	**0.653**	0.09	*
United Church	**1.115**	0.10	*
Anglican	1.029	0.68	
Presbyterian	0.987	0.91	
Lutheran	1.052	0.66	
Buddhist	0.819	0.29	
Jewish	0.838	0.28	
Ukrainian Catholic	1.073	0.74	
Religiosity			
Member of a religious affiliated group	**1.128**	0.03	*
Frequency of religious attendance	1.021	0.13	
Importance of religion	0.971	0.11	
* significant at 0.05 level or less			

Of these variables, the effects of sex and immigrant status happen to be statistically non-significant, which probably suggests that their effects on trust have not been inherent and authentic; rather, they have been mediated through the other variables in the model. The variables with statistically significant effects on trust are age, rural residence, income, and education. An older age and a rural residency, as hypothesized, result in more trust. The impact of income is negative, contrary to what the literature suggested. This, to some extent, speaks against those conceptualizations of trust that consider a cost-benefit analysis to be at the heart of the decision to trust. It should be noted, however, that in both cases – age and income – the impact is quite small and, therefore, hard to serve as the basis for a strong argument in either direction.

The demographic variable with a noticeable, consistent, and statistically significant impact on trust is education. The positive impact of education on increasing one's likelihood to trust has been confirmed in many other studies as well. According to the reported values of Exp(B), the lowest level of trust is found among those with less than a high-school diploma (which is the omitted reference category), but with the increase in years of education, the trust level rises systematically. This is good news for the policy-makers who might be concerned about finding ways to raise the trust level in Canada, as education is probably the most policy-friendly variable.

The second group of variables discussed above – i.e., marital status, social networks, volunteering, civic engagement, ties to local community, and the media – have also generated interesting results. With regard to marital status, a mix of expected and unexpected patterns surfaced. Being married (the omitted reference category), as was hypothesized, generates the highest level of trust compared to all other categories under marital status. Again, as expected, being divorced – and more than that, being separated – pushes trust levels down in a significant way. Also, being single or in a common-law relationship have a similarly negative, and equally strong, impact on trust. The fact that the latter two groups – singles and those in common-law relationships – have a lower level of trust is not unexpected; what is unusual is that their effects are so similar to those reported for divorced and separated individuals. Another curious phenomenon is that being widowed also generates a lower level of trust compared to being married, although to a lesser extent than being divorced or separated does.

This might have to do with a sense of vulnerability that widows, particularly at older ages, might feel. It can also be a product of a reduced level of social interactions after the death of a partner.

As expected, having a larger network of friends, and involvement in volunteer activities facilitate the development of trusting views. The type of civic associations that one is a member of also has some consequences for trust. Of all the different types of civic activities, only two are statistically significant. Of these two, membership in sports and recreational activities has a positive impact, and membership in fraternal organizations and service clubs has a negative impact on trust.

The nature of ties to local communities matters a lot for the development of trustful views. Here, those with statistically significant effects are the general sense of belonging to the local community and the opportunity to receive a favour from a neighbour, both with positive impacts on trust. The latter is particularly interesting, given its contrast with a closely-related variable in the model – i.e., whether or not the respondent has done a favour for a neighbour – which generated a non-significant negative impact. This indicates, understandably, that being on the receiving end of the caring gestures by others may go a long way in generating trust not only in those particular individuals but in the population at large. This is particularly important from a policy-making point of view, as it provides an effective policy tool in targeting the unusually low-trusting groups.

When it comes to the effect of media on trust, the pattern which surfaced seems to be along the lines of what Putnam suggested for the American society. The statistically significant variables with negative effects on trust are: following news and current affairs, using TV as the main source of news, and hours of watching TV. In contrast, using any other medium – newspapers, radio, and the Internet – is associated with a higher level of trust, with newspaper being the forerunner. The fact that following news and current affairs, controlling for the medium which is used, lowers the level of trust in others might be related to the fact that most of the events that find their way into the news are predominantly negative and worrisome and, therefore, might convey a pessimistic view of the world and the communities we live in.

The results for the impacts of religion-related variables show that, compared to those with no religion, all those who identify themselves with a particular religious background tend to have lower trust in others. This includes Roman Catholic, Baptist, Eastern Orthodox, Muslim, Hindu, Pentecostal, and Jehovah's Witnesses. Relaxing the threshold for statistical significance from 95% to 90% adds Sikhs to the above list, but it also brings in the United Church with a positive impact on trust.

Having a religious background is not the same as being an affiliated and/or active member of the faith community. While the former is merely an abstract identity marker, the latter involves social interactions with others. Interestingly enough, being an affiliated and active member of a faith community has a totally different effect on trust than that of religious background; it raises the trust in others, and this effect is statistically significant.

The above contrast in the effects of religious background as a mere identity marker and religious involvement as social interaction also resurfaces when we examine the effects of two other variables: the importance of religious beliefs, and the frequency of religious attendance. Interestingly enough, while the former lowers trust, the latter raises it; although, in order to get statistically significant effects for these two variables, we have to lower the confidence level to 87% (which is a legitimate thing to do, given the theoretical purposes of our study). What is most important and interesting is the contrast between the effects of the two variables and the similarity of their effects with those of the religious background and religious affiliation variables; in both pairs, it is the variable involving social interaction that raises the level of trust in others.

Conclusion

The simultaneous inclusion of the variables affecting an individual's likelihood to trust others enormously enriches the insight developed through the previous chapters. The major benefit of using logistic regression with these variables was the ability to control for the effects of all other variables, hence, the ability to see the pure impact of each variable. It also allowed for a more refined picture on the effects of the various religion-related variables.

A general theme that seems to be surfacing here as well is that the variables which are apt to increase the likelihood of trust tend to be those that promote or involve social interactions. The suggested parallel between the level of social interactions and the likelihood of trust could be seen most visibly in the case of religion. As it became clear above, the religious features that act mostly as identity markers seem to be functioning as separating lines between individuals and others, and lowering the trust as a result. On the other hand, those religious features that involve some interaction with others had an entirely opposite effect. Some of the previous studies on the relationship between religion and trust – which had indicated a lower level of trust among those affiliated with hierarchically structured religious communities (Schoenfeld, 1978; Putnam, 1993; La Porta et al., 1997) – may indeed have been speaking to a similar phenomenon; vertically-structured religious organizations provide less of an opportunity for the members to interact with each other on an equal footing.

Some of the surprising findings in regards to the impacts of various marital statuses on trust could also be better explained using the social interaction conjecture. In the findings, we noted that the highest level of trust could be found among married individuals, and the lowest among the divorced, separated, single and, to a lesser extent, those in common-law relationships. What was somewhat perplexing was the lowering effects of being single or widowed. Viewing these two statuses from a social interaction angle, however, would indicate that these two groups also have fewer or less intense social interactions with others, and that their trust level may be to some extent associated with that reduced interaction.

A similar theme can be found for the variables involving local community and neighbourhood features. The fact that a neighbour has done a favour for the individual goes a long way in raising his trust level in others. Interestingly, neither the number of neighbours one knows, nor the fact that he or she has done a favour for a neighbour, has a significant impact on one's level of trust. By contrast, the size of one's social network and also volunteering – both clearly associated with a higher level of interaction – raise the trust levels, as does living in a rural area.

The reverse side of the story – i.e., the variables that failed to clear the statistical significance bar – is also equally telling. Among these variables were sex and immigrant status. It seems that after controlling for the effects of the rest of the variables, there is no particular impact which can be attributed to these two; this makes sense given the above discussions, as neither of these variables, by itself, indicates a particular level of social interaction. The dampening impact of income on trust might also be a part of the same phenomenon, if it can be shown that a higher income would mean more independence and, as a result, less interaction with others.

The results of the logistic regression discussed in this chapter adds another necessary piece of information – i.e., correlates of trust at the individual level – to the picture acquired in previous chapters. The survey data used in the process of logistic regression analysis, however, like any other quantitative data, does a better job of covering a large surface than going into the depth of an issue. Capturing the latter is only possible through the use of qualitative methods. In the next chapter, we report a host of supplementary findings based on a series of interviews.

PART IV:

Putting things together

13 The Voices Behind the Statistics

The analyses offered in previous chapters were based mostly on survey data. This kind of data allows for grasping general trends and big pictures, but cannot easily capture phenomena at deeper levels. They are inherently limited in terms of how much they can reveal about the things happening behind the scenes and the mechanisms through which different positions and opinions are arrived at. To capture these deeper layers, which are present in any issue worthy of sociological investigation, we need to subscribe to qualitative methods, of which the most commonly used is in-depth interviewing. In this chapter, we discuss the major findings that emerged from long interviews conducted with seven immigrants. While the typically small number of people involved in qualitative research does not warrant a quick generalization of the findings, the resultant information sometimes offers invaluable insights. Such knowledge can be used either in making better sense of the patterns which surfaced through quantitative data, or they can act as a basis for developing more sophisticated hypotheses for future research.

One of the key issues on which these interviews helped shed some light was the complex role that culture plays in immigrants' social interaction with the native-born population. The different cultures make it very difficult for the inter-cultural communication of feelings and emotions. One immigrant, a woman from Peru, who has been in Canada for more than ten years, expresses this difficulty clearly:

> [back home] we greet people with kiss and hug; I just came and my husband introduced me to people, and I hugged them and kissed them; and they were like, taking some space; I felt very bad about that; …. They felt weird, and I felt weird, too ….. Now, I am like, hi, hello, that's it…. Also, there, we didn't have to set appointment to go to each other's houses; we just stop by…

The ability to predict the feelings of other parties involved in an interaction creates a more comfortable environment. But that should not be confused with trust. In other words, people can feel more comfortable with some others, without necessarily trusting them more. When asked about which ethnic groups and/or nationalities could be trusted more easily, one immigrant woman from Colombia said:

> Latin/Central/South Americans! ... yes, I think I can rate them better than any of the other groups. I don't know if I can say that I trust them, but I feel more at ease with myself, because I know how to read them better...

Another immigrant, from Iran, commented on her biggest challenges after migrating to Canada and the reasons for her small network of friends:

> [the biggest challenge was] culture; I could not understand people. Part of that was because of language; so, I had a hard time. My husband always tells me why you don't find a friend there [at the university where I study]; it was only this last semester, after four years, that it became better. Students talk about things – movies, characters, etc., – that we could not understand....

When asked about the ethnic groups and/or nationalities in whom she has more trust, the same person commented:

> ... [with] Iranians, you have to be really careful, I guess; it is kind of hard to have relationship with them, because they come from different backgrounds; when we came here, we thought we could hang out with all Iranians; but, then we learned we didn't have much in common.

How can these two seemingly opposite views be reconciled? How is it that an immigrant cannot connect socially with the mainstream population, but at the same time, express little trust in his or her compatriots, either? The answer seems to lie in the fact that a common culture does not necessarily push people towards liking each other more, but rather towards understanding one another better. Cultural commonality seems to inject more predictability in social interactions and, through this, provides a necessary framework within which actions can be understood more clearly and the motives behind them read more effectively. In a common cultural domain, the expectations can be known with much more clarity and, therefore, the anxiety associated with ambiguity may diminish significantly. In other words, a common culture makes social interactions more economical, as it saves a great deal of so-called emotional expense that would have otherwise been associated with the relationship.

This is an important point with implications for the research on trust. Predictability of actions and reactions does not necessarily lead to trust. Confusion, ambiguity, and lack of certainty about what to expect is one thing; knowing and disliking this expected behaviour is another. Both generate frustration, but the latter kind of frustration is different from the former. While one is a product of not knowing what to expect, the other is the result of different statuses within the parameters of known expectations.

The above point is reflected, to some extent, in the statement made by an immigrant from Scandinavia, who had expressed distrust towards eastern Europeans and those from the former Soviet Union countries, except for the Polish:

> ... because it [there] is a Catholic influence there – with that comes some ethics that may lack in some of the former Soviet Union countries.

The fact that this is said by someone who was brought up Lutheran may be taken to imply that it is not the religion itself, but the predictability of behaviours influenced by a known ethical system, that matters. Despite the difference in faiths, the presence of the Catholic faith among the Polish meant, for this immigrant, more stability and clarity in their actions.

The impact of religion, however, is not limited to the generation of more predictability. Another reason why the presence of faith and religiosity can make a difference might have to do with the presumed presence of a certain degree of selflessness and honesty on the part of religious individuals. These two, as opposed to self-interestedness, arrogance, and deception, can go a long way in building trusting relationships. This theme was raised very frequently during the interviews. The immigrant from Colombia, when expressing the reasons for his distrust towards those of Italian and American origins, said:

> I would be hesitant [to trust Italians]; I don't know why, but I have this perception that they have a double agenda.... Americans? I don't really trust them; people that I have interacted with have a cockiness, and think too highly of themselves, so I don't feel very good about them.

The female immigrant from Scandinavia had something to say about Americans, which was more or less along the same lines:

> Americans as a group can be annoying, but as individuals, they can be really nice and could be trusted.

The influence of culture on shaping the social ties between people goes beyond the aspects mentioned so far. Indeed, some of the other factors that are perceived to have suppressed social capital are either proxies for culture, or function through it. Take, for instance, the influence of TV, the pressure of time, and the pressure of money, as three main culprits in Putnam's view, for keeping the American social capital stock down. As far as TV goes, Putnam believes that it has resulted in passive and privatized entertainment and therefore has pushed the more interactive and collective forms of entertainment into a corner. This was corroborated by our interviewees, but one of them also pointed to another aspect in which TV has influenced our social ties, that is, through creating a 'culture of fear':

> That's something that I realized when we had cable TV where we used to live… We got the news from the States and they had a lot of fear-mongering on the television. We were watching more TV at the time than now. I had more fear at the time when I was watching true stories about people murdering people , etc., …

There exists a strong cultural element in the way in which the pressure of time works. The immigrant from Iran was quick to agree that the pressure of time exists and is real:

> I guess people here are really busy; they don't really have time for other people.

But when she commented further it became clear that, for her, this was not necessarily a natural state of affairs, and involved a choice on the part of Canadians as much as it was the imposition of the hastiness of the modern lifestyle:

> … this is one of the things I always complain … because I always say, we are friends, but you don't have enough time for me; so, which kind of friend ….? Because, in Iran, we always make time for each other. You know, we just have one day off in

> Iran, the Fridays, but always in that day, we meet our friends, see our family; but people here are so busy, so tired, they are with their family, not really with their friends, or each other. Or, in weekend, they just want to relax; no friends.....[in Iran] we always have time for each other. I really always complain about this; and when I tell Canadians, they say, "we are just busy, you tell us, we make time for you"; no, I never ask you to make time for me; you should do that without I tell you. But, in Iran, we just do that; even in bigger cities, longer distance, we always have time for each other; at least, we call each other.

The literature cited in previous chapters is far from conclusive about the nature of the relationships between money and social connectedness. Some, like Wuthnow (2002), have attributed the declining American social capital to the drop in the resources available to people. The opposite side of his argument is that of Putnam (2000), who considers the pressure of acquiring more money as a causative factor in this process; for him, it is not the increased poverty, but the wanting of more, that has killed the civic-mindedness of Americans. On the other hand, looking at the issue of trust from a rational choice perspective, Hardin (1999) argues that, because of the risks associated with trusting unknown others, those with higher income should be better prepared for the absorption of the costs of such risks and, hence, they should be more trusting. Drawing upon their international experiences, the immigrants interviewed offer some interesting observations, which could be useful in deepening our understanding of how money and social connectedness interact.

Spending a lot of days and nights in backpackers' hostels, the immigrant from Colombia noticed, to his surprise, no report of anything stolen or taken in the hostel:

> I'm pretty impressed that there's not too many cases of people having their stuff taken. At first, I wouldn't leave anything on the bed, if I had to go to washroom or something... I would make sure everything was locked up. After the second week, I would leave my cell phone on my bed, because I could see that other people didn't have a problem with that.

The immigrant from Scandinavia, who has apparently been a frequent traveler, mentioned something more interesting along these lines, that in most of her trips, particularly in poorer countries, she had stayed in people's houses rather than hotels:

> Poorer people are more likely to invite you to stay. [In] poor countries, people are more willing to share the little resources they have than we are… In Brazil, we stayed, out of one whole month, for only one week in a hotel; the rest we stayed with people. We met people on the airplane and some in a hotel – he had a brother in Rio de Janeiro, who we stayed with. He had a great nephew or something in another village; we just told this old man where we were going, and he knew people there… Also, when I was hitch-hiking, the people who stopped, they didn't necessarily have the nicest cars…

As for reasons as to why poor people are more willing to share, she offered the following insight:

> They can see the need a little more – if they were traveling, would they be able to afford a hotel? … Maybe, also, because if you don't have lots of valuables in your house, it may not be too expensive to replace – I don't know though.

The above reasoning was, to some extent, echoed in one quantitative piece of information we discussed in chapter seven. As shown in Figure 7.4, the relationship between the average income of city dwellers and their overall level of trust in others had a curvilinear pattern, that is, the trust level rises as income increases, but beyond a certain point it stops and then starts dropping. One possible explanation for the curvilinear nature of the relationships is that after the income increase continues beyond a certain point, the psyche and cultural outlook of people may start to change, making them more conservative in order to protect their possessions. In order for people to share their resources with others, those resources need to reach a critical mass to allow accommodation of others; but beyond a certain point, the replacement of things that might be lost as a result of trusting others might get too costly, forcing people to become more protective of their belongings. In either case, the major implication is that the relationship between income and various aspects of social capital is far from straightforward, and much more complex than the initial conceptualizations of the issue have offered.

Another perhaps equally surprising phenomenon that Putnam found in his study was the fact that married individuals, and particularly those with children, are the most civically active members of society. This sounds somewhat paradoxical, as these individuals are those with the busiest lives, least amount of time, and the highest pressure of money.

While he does not offer a good explanation for why this is the case, he points to some of correlates that might be the causes: those with families are also most likely to be employed, involved with the schools of their children, engaged in volunteer activities related to their children's participation in sports, and so on. In all these cases, children seem to be a focal point in boosting the civic life of their parents. The interviewees of our study also confirm the above general conjecture in more than one way.

The immigrant from Scandinavia pointed to subtle ways in which the presence of children can make parents more socially engaged. Reflecting on the scarcity of contacts with neighbours in her teen years, for instance, she said:

> I grew up in ... a small village, and then, my parents moved to a bigger place, and there was a little bit less [contact with neighbours] there. [Partly] because I didn't really live there [long enough], but also because my parents didn't have kids anymore, and that's a really great way to meet neighbours, because you walk more, you are outside more.

In contrast to the above, however, her own children are causing her to be more socially engaged, though in some indirect ways. Pointing to the creative way in which she and several other women formed their own little 'clubs' –e.g., babysitting club, cooking club, etc. – she described the reasons behind the formation of such groups in the following way:

> ...we [the couple] do need time off [from childcare responsibilities]... [but] we can't just drop off the kids at a babysitter at any time... Through this group [a group of women] and other friends, we organized a babysitting club where two ladies get together for one day – three hours in the morning – watch all the kids (12 of them), and then they're a couple of sessions off ... Last time, we got one babysitting session for every two times off... We'll put the kids to bed and then we'll go out... we do that once a week...

The goals, and also the means to achieve the goals, are cleverly articulated by this woman, capturing the essence of what social capital does. The goal was to satisfy a need: the need of parents to have some time off from childcare responsibilities; and, to satisfy this need, as the interviewee put it, "we've got time, but no money." So, the smart strategy was to convert the time into money, through forming their babysitting club.

This not only highlights the fact that social capital has functional relevance, but also illustrates the notion of fungibility of different types of capital that Bourdieu tried to emphasize through his 'economy of social practices.'

The formation of the 'cooking club' was not inspired by the same economic concerns; it was more to satisfy the need to be socially engaged. This is how she put it:

> One of the people in the babysitting club got the idea to cook together – because we like to be together... It's kind of unnatural to just stay in your own house and be lonely all day and be alone with your own kids. So she said, "Why don't we get together and cook a few casseroles..." We'll just get together and cook up a lot of food and then divide it. It's a good idea because you don't have to stay at home and cook; and, the kids have fun together, too... I think that the social is the main reason to do it, and the economic reasons are secondary.

Ironically, the motivation to form groups like the above cooking club is sometimes fed by not the absence of economic resources but the excess of it. Commenting on the reasons why people are less socially connected to one another today than before, the same immigrant pointed to the effect of the so called 'separate economies:'

> Now you have your own car ... and you probably don't need so much help – you can make it on your own. So, before, it was more practical [to help each other]....

In order to bear fruit, this need to connect socially has to be accompanied by a general tendency to trust others; and, most of our interviewees had no hesitation to express this tendency. The students of trust scholarship, however, are divided between those who believe the tendency to trust is one triggered by cultural orientations and/or personality traits (Offe, 1999), versus those who believe it is shaped by situational factors, that is, a rational analysis of the costs and benefits of trusting others, after every dramatic social experience involving trust (Hardin, 1999). The information from the interviews we conducted argue in favour of a third possibility, which makes the above either-or dichotomy unnecessary; that is, people are driven by a cultural and/or personality tendency – mostly a positive and trusting view – but this view gets modified along the way by their personal experiences, particularly when one experiences the feeling that one's honesty and trusting character has been taken advantage of.

On the issue of whether or not people can be generally trusted, the immigrant from Iran had this to say:

> I normally trust people, at the beginning; then I continue my relationship; and I'm really honest in my relationship; ... but, after sometime, if we see something, we just, you know, get back, and just say, no relationship.... If they take advantage of me, I just cut my relationship... [with some friends] we felt that the way we are honest with them, they are not... But, I sometimes think if I don't start with that honesty and just be more careful, it doesn't happen.

A woman in her fifties, who has been living in Canada for twenty-some years, echoed similar feelings:

> I always think that people can be trusted – people say I'm too much of a sucker. I always believe good of everyone and not bad. We just had a guy at work who turned out to be bad, it turns out that he had done some really horrible things, he's left now.... I was totally suckered because I thought he was a nice guy. He always seemed to work hard – he seemed nice.... [Sometimes] people would take advantage of you, and make you do things that they know you won't say no.... [Now] I've learned to say no; they were taking advantage...

Corroborating the same general point, a woman who migrated to Canada from Greenland, talked about her experiences the same way:

> When I moved from [my town in Greenland]... to Calgary, I had the Northern mentality. I didn't know how people actually were like in the southern part of Canada; I really didn't. I came to Calgary, I guess you could say that I was very naïve, very trusting, believing everyone, very trusting, that sort of thing... what I'm trying to say is that you can trust people in the North – with your children, [for instance] – while you can't really do that here. I had to change my thinking in that way.
> In Greenland, you can have babies in the stroller, and if you go into the store, you can actually leave a baby outside of the store, and they are completely safe. If they are crying, passers-by will rock the stroller, until the mother comes back. If the baby goes to sleep, that is fine. In Calgary, I actually left one of my children in the vehicle, when I went into an office. I wasn't thinking, and I got a big slap on my hand for it. Social workers phoned and asked me why I did it, and what happened. That was a learning experience for me. I just basically told her that I wasn't thinking, I was in the process of moving .... and I explained that to her, and she was fine with it.

Conclusion

Some researchers and policy-makers concerned with the integration of immigrants into host societies can be quick to judge the strong connections between immigrants and their ethnic communities as a counterproductive experience, as a factor that slows down their integration into mainstream society. Such researchers are right in their conclusion that in order for immigrants to become fully functional members of their new homes they need to develop ties with the society at large and not just with their ethnic/cultural communities. However, in coming to this conclusion they ignore or downplay two things: the barriers to full integration posed by the society at large, and the vital roles that such communities play in the well-being of immigrants.

As is clear from our discussions in previous chapters, the formation of effective social capital is heavily dependent upon the presence of a favourable cultural and normative environment. The cultural gaps between immigrants and the mainstream population are too wide to be bridged, at the least in the period immediately after arrival, through artificial, quick-fix interventions from outside. During this initial period, the encounters between immigrants and non-immigrants can be charged with a lot of unavoidable misunderstandings. Only time, and the accumulation of mutual knowledge through increased interaction, can remedy this problem.

Even when the above barriers are not in play, there will remain another force that ties immigrants to their ethnic/cultural communities, that is, the roles and functions of those communities. The reality is that, even in the most industrial and individualistic societies, people rarely live their lives fully independently, without any help from their surrounding communities, whether this means their extended family, their circle of friends, or more formal organizations. In the lives of immigrants, that whole community component is missing and has to be replaced with something else, and the closest entity to that missing component is their ethnic communities. The following passage is what the immigrant from Peru had this to say about this:

> I tried to be ... with the Latin people; we have lots of friends from Nicaragua, Venezuela, El-Salvador, Mexico... we help each other ... When I had my baby and my parents couldn't come, they took my daughter out; they took care of my daughter; and I try to do the same thing when they need help...

In sum, there are many things – e.g., teaching the language of the home country to the children of the immigrants, providing emotional support to them in times of difficulty, assuaging the sense of loneliness, etc. – that the society at large cannot provide to immigrants, at least not for a while. For all those reasons, the attempts to theorize the integration of immigrants needs to be more humble and humanitarian, or else it runs the risk of being naïve and irrelevant.

14 Wrap-up: Conclusions and Implications

By now, our discussions in this book should have given some indication that, when it comes to social capital, Canada provides a unique example. The highlights of this potentially unique Canadian story are as follows. First, Canada has not experienced the huge and worrisome drop in social capital that countries such as the United States or France have witnessed. To this, we need to immediately add a qualifying term, that is, the Canadian data that we have utilized here cover only the past two decades or so, while the American studies are based on data from the 1960s onward. Second, during the last two decades of the 20th century, the early 1990s acts as a watershed showing the peak of many Canadian trends, particularly those related to political participation. Third, this particular trajectory is almost identical to the one found for Germany. The search for factors behind this incredible resemblance is beyond the scope of the present study, but it is interesting to note that each of these two countries, in their own way, has experienced a political split within: Quebec versus the rest of Canada, and West versus East Germany. How this split has influenced social capital trends could be a fascinating topic for future research. Fourth, the relationship between ethnic diversity and trust – which was negative in most of the major studies done on other countries – turned out to be positive in Canada.

There also exists some interesting within-Canada variations, the most visible of which involves the distinct status of Quebec. Compared to all other provinces, Quebec shows a noticeably lower level of trust, volunteering, and social connectedness, as well as a higher level of confidence in private and public institutions. Also, within Quebec, Montreal has a unique path of its own as far as the relationship between ethnic diversity and trust goes. While the overall pattern in Quebec indicates a strong positive relationship between these two variables, Montreal proves to be an anomaly, combining a high level of diversity with a low level of trust. The latter point does also make for a fascinating new line of future research.

In an initial attempt to uncover some of the forces behind the unique status of Montreal, we arrived at one interesting potential culprit, namely, the residential segregation of people of various ethnic backgrounds. Indeed, the level of trust expressed by members of each ethnic/cultural group was inversely related to their average level of spatial segregation from all the rest of groups in the city; the least segregated ones were the most trusting, and vice versa. To conceptualize the latter point, we hypothesized that one's trust in other people – particularly, in unknown others – is directly related to the intensity of their interactions with people of different backgrounds; the presence of residential segregation, therefore, could lower the level of trust for a group through limiting the number of their interactions with people who are different from them.

The above point surfaced again in a different context, i.e., the level of trust expressed by different generations of immigrants. Interestingly enough, the immigrants coming from a more diverse ancestry – indicated by the birthplaces of their parents and grandparents – had a higher tendency to trust. In contrast, those born to relatively homogeneous ancestors – either all Canadian-born or all foreign-born – happened to be the least trusting ones. This speaks to the significance of diversity – or, exposure to diversity to be more accurate – in influencing the ways in which we connect to others.

Before going any further, let's restate the above point a little more clearly! In Canada, the cities that are more ethnically diverse have shown a higher level of trust (with the exception of Montreal, of course). Within cities, the ethnic/cultural groups that are less segregated from other groups have also shown a higher degree of trust (with few exceptions). Among the population, those with a more culturally diverse ancestry report the highest level of trust in others. In other words, a higher trust in unknown others is associated with having interacted with, or having been exposed to, different others. How can this particular type of connection between trust and exposure to diversity be explained?

Unfortunately, the still very recent literature on social capital and trust has not offered much that can be of help here, partly because the positive relationship between trust and diversity has rarely surfaced in other studies. Also, most of this literature has been generated by political scientists, sociologists, and economists, who have been more concerned with macro-trends than inter-personal interactions. In our search for a useful theoretical framework for conceptualizing of the above issue, we came across a long stream of research in social psychology which seems to be capable of offering some guidelines in this regard. This research has been primarily formed by social psychologists concerned with the relationships between different racial groups. One of the biggest topics of interest to these researchers has been the interplay of social interaction and prejudice; that is, whether or not the social interaction of individuals with different racial origins could result in a change in their initial racial prejudices and stereotypes. The experimental studies conducted by these researchers have generated a distinct body of literature that is now known as the 'contact theory.' Below, we outline the main tenets of this stream of research and, then, we discuss how this might help us understand the Canadian situation.

Contact Theory

The first studies in the research stream that came to be known as 'contact theory' were conducted in post-WWII America, the most influential of which was a study by Allport (1979 [1954]) entitled *The Nature of Prejudice*. In this work, Allport showed that contacts between racial groups can have the positive impact of shattering their initial prejudices and transforming their stereotypes towards each other. For this positive impact to come about he thought that four key conditions needed to be present: the groups in contact should be of equal status; the contacts should happen during a process of collaboration, and not competition; a specific common goal needs to be attached to this process; and, finally, all these have to happen within a supportive environment in terms of authorities, laws, or customs (Pettigrew, 1998).

The ways in which inter-group contacts affect prejudices, according to Pettigrew (1998), consist of four potential mechanisms. The first mechanism involves learning about the outgroup; through contacts the members of each group accumulate more information about the other group and, in the process, fix their superficially-constructed stereotypes.

The second mechanism has to do with a change of behaviour among the members of the groups involved; the cooperation between the members of different groups for the achievement of the common goals would create an imbalance between their behaviours and their initial prejudices, and since they cannot change the former, the only way to remove this imbalance is to change the latter. The third process revolves around the emotional and affective ties that are developed, in the process of collaboration and continued contacts, between individuals belonging to different groups; the emotional bonds exert a huge pressure on the initial prejudices in order to secure consistency between the two. The fourth mechanism involves a bigger and more fundamental change in the ways in which each group views the social world; now that each of them has been exposed to a different lifestyle and culture, they are more likely to shed their own narrow and ethnocentric views of life in favour of becoming more tolerant of difference.

Despite some mixed results, the overwhelming majority of the studies done to test the validity of the 'contact hypothesis' have been supportive of its basic tenets. In a recent meta-analytic survey of all the existing studies that are related to the contact theory one way or another, Pettigrew and Tropp (2006) found consistent support for this theory, regardless of the location of the studies, their sample sizes, and also the context within which the contacts have been made. This latter point is, in and of itself, a major finding, as it allows contact theory to transcend its initial, and still ongoing, focus on interactions between neighbours (Powers and Ellison, 1995) to many more diverse settings in which interactions could take place, such as contacts among students (Moody, 2001), roommates (Laar et. al., 2005), and co-workers, as well as free-floating interactions (Sigelman and Welch, 1993; Emerson et. al., 2002), casual conversations (Sigelman et. al., 1996), and contacts among church attendants (Yancey, 1999; DeYoung et. al., 2005). The confirmation of the contact conjecture in such diverse settings has been crucial in upgrading its status from being a 'hypothesis' to a 'theory.'

It is this wide applicability of the contact theory that makes it relevant for our study of trust here. Trust, in essence, is nothing but a prejudicial state of mind towards 'others.' In the absence of any concrete information about 'others,' and in the absence of any prior contact with them, one has to rely on some casual observations, superficial information, and views transmitted from previous generations in order to decide about the extent to which he or she can engage in a serious relationship with the unknown other. Like all other negative prejudices, a tendency to distrust can also shatter as a result of repeated contacts leading to increased knowledge of others, development of affective ties, change of behaviour, and the weakening of ethnocentric views. Such mechanisms could be at work in all the situations discussed above, in which a higher level of trust was found to be associated with more exposure to diversity: at the city level, the neighbourhood level, and at the immigrant generational level.

The argument suggested above for a relationship between the potential for interaction and trust is not without precedent. Coleman (1990) used the same idea to explain an entirely different phenomenon, that is, the rise of the motto, "Don't trust anyone over 30," among the youth in 1960s and 1970s America. According to him, the popularity of this motto had to do with the demographic composition of the American society during that time, in which the baby boomers had reached their young adult years, increasing the ratio of youngsters to adults in the population. As a result, an average young person experienced "an increased proportion of communications from those of the same age (p: 191)." By the same token, he suggested the following for a possible relationship between ethnic composition and trust:

> [A]s ethnic composition of population changes, unless association is tightly enclosed within ethnic groups, the ethnic distribution of persons in whose judgment trust might be placed changes for each member of the population. (Coleman, 1990:192)

There is one more question to address: if contact and exposure to diversity can have such an incredible effect in corroding distrustful views, why is this the case only in Canada? Two possibilities can be entertained here, although both remain highly theoretical and neither are amenable to rigorous empirical verification, at least not with the existing data. The first possible explanation involves the degree of diversity in Canadian society. There are many other immigrant-receiving countries in the world, and some, like the United States, accept a much larger number of immigrants annually than Canada does. However, the Canadian intake of immigrants per year is higher than any other country in proportion to its population size. No other country in the world admits close to one percent of its population from abroad every year, as Canada does.

How can this higher level of diversity proliferate trust among members of a society? This question is all the more serious for those who believe that the presence of minorities has the potential to trigger a negative and hostile view towards them on the part of the majority. Indeed, there exists a strand of research on inter-group relations, under the title of 'group threat theory,' the basic tenet of which is that the increase in the population of minority groups could be perceived by the majority as threatening – politically, economically, and culturally – and can result in more hostile and less trusting views between groups (Bonacich, 1979; Lieberson, 1981). However, Blalock (1967), one of the supporters of this thesis, argues that the relationship between these two is not linear, but curvilinear. That is, the increase in the population of minorities triggers a negative reaction, but only up to a certain point, beyond which the attitudes start to change in the opposite direction. This is to suggest that there might exist a 'critical mass' point, beyond which diversity will turn into a trust-boosting force. The level of ethnic and cultural diversity in the Canadian population may have passed this critical point, while that of other countries has not.

The second factor that could have potentially contributed in creating a positive relationship between exposure to diversity and trust has to do with the impacts of the nation's official policy of Multiculturalism. Here again, the contact theory literature might be of some help. As mentioned above, according to Allport (1979[1954]), one of the conditions under which contact can bear positive fruits is the presence of a supportive legal and normative environment. The celebration of cultural and ethnic diversity in Canada could have been a factor creating the supportive environment needed in order for the positive relationship between diversity and trust to take shape. Indeed, some signs of a fundamental difference between Canada and other immigrant-receiving countries in this regard have already surfaced, the most noteworthy of which appeared at the time of the global rage among Muslim minorities in many European countries against the publication of the cartoons of the Prophet Mohammad in a Danish magazine. The almost total absence of massive and outraged demonstrations by Muslims in Canada came as a surprise to many observers; one of whom argued, in an article in *The Globe and*

Mail, that "[m]ulticulturalism and media likely muted protests (Valpy, 2006)."

The latter point about the potential impact of Multiculturalism needs be taken up in a more serious way in future research. This is a crucial task, not only from the academic research point of view, but also from a policy-making perspective. Given the emerging discourse on post-Multiculturalism in many industrial nations, mostly in reaction to the recent rise of radicalism and support for fundamentalism among some minority groups, it is important to establish whether Canadian society is in the same boat in this regard with all those other nations.

A concern with the impacts of diversity in today's world will inevitably involve identity. Most of the challenges posed by diversity, and most of the achievements made in overcoming those challenges seem to have been working through the dynamics of identity formation mechanisms. This may partly explain the rise of interest in research on identity over the past decade or so, but it also calls for a better integration of the two streams of research.

References

Alesina, Alberto, and Eliana La Ferrara. 2000. "Participation in Heterogeneous Communities." *The Quarterly Journal of Economics*. August: 847–904.

Alesina, A., and Eliana La Ferrara. 2002. "Who trusts others?" *Journal of Public Economics*. 85: 307–334.

Allport, Gordon W. 1979 [1954]. *The Nature of Prejudice*. Garden City, NJ: Doubleday.

Aronson, Elliot. 1992. *The Social Animal* (6th edition). New York: W.H. Freeman and Company.

Becker, Gary. 1964. *Human Capital*. New York: Columbia University Press.

Blalock, Hubert M. Jr. 1967. *Toward a Theory of Minority-Group Relations*. New York: John Wiley & Sons, Inc.

Bonacich, Edna. 1979. "The Past, Present, and Future of Split Labour Market Theory." *Research in Race and Ethnic Relations*. 1: 17–64.

Bourdieu, Pierre. 2001 [1983]. "Forms of Capital." *The Sociology of Economic Life*. M. Granovetter and R. Swedberg, editors. Oxford: Westview Press. Pp. 96–111.

Bourdieu, Pierre, and Loic J. D. Wacquant. 1992. *An Invitation to Reflexive Sociology*. Chicago: University of Chicago Press.

CBC (Canadian Broadcasting Corporation). 2006. "Separation Anxiety: The 1995 Quebec Referendum." *The CBC Digital Archives Website*. Canadian Broadcasting Corporation. <http://archives.cbc.ca/IDD-1-73-1891/politics_economy/1995_referendum/>. Last updated: 20 March 2006. Accessed 18 July 2006.

Citizenship and Immigration Canada. 2006. *Final Report: G8 Experts Roundtable on Diversity and Integration*. October 4. Lisbon, Portugal.

Coffe, Hilde, and Benny Geys. 2006. "Community Heterogeneity: A Burden for the Creation of Social Capital?" *Social Science Quarterly*. 87(1): 1053–1072.

Coleman, James S. 1988. "Social Capital in the Creation of Human Capital." *American Journal of Sociology*. 94 (supplement): S95–S120.

Coleman, James S. 1990. *Foundations of Social Theory*. Cambridge: Harvard University Press.

DeYoung, Curtis Paul, Michael O. Emerson, George Yancey, and Karen Chai Kim. 2005. "All Churches Should be Multiracial: The Biblical Case." *Christianity Today* April: 32–35.

Duncan, Otis Dudley, and Beverly Duncan. 1955. "A Methodological Analysis of Segregation Analysis." *American Sociological Review.* 20(2): 210–217.

Durkheim, Emile. 1957. *Professional Ethics and Civic Morals.* New York: Routledge.

Durkheim, Emile. 1982. *The Rules of Sociological Method.* New York: The Free Press.

Durkheim, Emile. 1984 [1893]. *The Division of Labor in Society.* New York: The Free Press.

Emerson, Michael O., Rachel Tolbert Kimbro, and George Yancey. 2002. "Contact Theory Extended: The Effects of Prior Racial Contact on Current Social Ties." *Social Science Quarterly.* 83(3): 745–761.

Fernandez-Kelly, Patricia M. 1995. "Social and Cultural Capital in the Urban Ghetto: Implications for the Economic Sociology of Immigration." Pp. 213–247. *Economic Sociology of Immigration.* Portes, editor. New York: Russell Sage Foundations.

Frankfort-Nachmias, Chava, and Anna Leon-Guerrero. 2006. *Social Statistics for a Diverse Society* (4th edition). New York: Pine Forge.

Fukuyama, Francis. 1995a. "Social Capital and the Global Economy." *Foreign Affairs.* 74(5): 89–103.

Fukuyama, Francis. 1995b. *Trust: The Social Virtues and the Creation of Prosperity.* New York: Simon and Schuster.

Gambetta, Diego. 1988. "Mafia: The Price of Distrust." Pp. 158–175. *Trust: Making and Breaking Cooperative Relations.* D. Gambetta, editor. Oxford: Blackwell.

Gambetta, Diego. 1993. *The Sicilian Mafia: The Business of Private Protection.* Cambridge: Harvard University Press.

Goodhart, David. 2004. "Too Diverse?" *Prospect.* Feburary: 30–37.

Geertz, Clifford. 1963. *Peddlers and Princes.* Chicago: University of Chicago Press.

Granovetter, Mark., and Richard Swedberg, editors. 2001. *The Sociology of Economic Life* (2nd edition). Cambridge: Westview Press.

Halli, S. S., and K. V. Rao. 1992. *Advanced Techniques of Population Analysis.* New York: Plenum Press.

Hardin, Russell. 1999. "Do We Want Trust in Government?" Pp. 22–41. *Democracy and Trust.* M. Warren, editor. Cambridge: Cambridge University Press.

Hardin, Russell. 2002. *Trust and Trustworthiness.* New York: Russel Sage Foundation Publications.

Helliwell, John. F. 1996a. "Do Borders Matter for Social Capital? Economic Growth and Civic Culture in U.S. States and Canadian Provinces." *Working Paper no.5863*. National Bureau of Economic Research. http://papers.nber.org/papers/w5863

Helliwell, John F. 1996b. "Do Borders Matter For Social Capital? Economic Growth and Civic Culture in U.S. States and Canadian Provinces." *Working Paper no.5863*. NBER. Pp. 19-42. (Cambridge: National Bureau of Economic Research). Subsequently published in *The Economic Implications of Social Cohesion*. 2003. Lars Osberg, editor. Toronto: University of Toronto Press.

Honore, Carl. 2004. *In Praise of Slow*. Toronto: Vantage Canada.

Huntington, Samuel. 1996. *The Clash of Civilizations and the Remaking of World Order*. New York: Touchstone.

Ingelhart, Ronald. 1999. "Trust, Well-Being, and Democracy." Pp. 88–120. *Democracy and Trust*. M. Warren, editor. Cambridge: Cambridge University Press.

Ingelhart, Ronald, and Pippa Norris. 2003. *Rising Tide: Gender Inequality and Cultural Change around the World*. Cambridge: Cambridge University Press.

Isajiw, Wsevolod W. 1999. *Understanding Diversity: Ethnicity and Race in the Canadian Context*. Toronto: Thompson Educational Publishing, Inc.

James, David R., and Karl E. Taeuber. 1985. "Measures of Segregation." Pp. 1–31. *Sociological Methodology*. N.B. Tuma, editor. San Francisco: Jossey-Bass.

Kay, Fiona M., and Paul Bernard. 2007. "The Dynamics of Social Capital: Who Wants to Stay In If Nobody Is Out?" Pp.41–66. Social Capital, Diversity, and the Welfare State. Kay and Johnston, editors. Vancouver: University of British Columbia Press.

Kay, Fiona M., and Richard Johnston. 2007. *Social Capital, Diversity, and the Welfare State*. Vancouver: UBC press.

Kazemipur, Abdolmohammad. 2000. "The Ecology of Deprivation: Spatial Concentration of Poverty in Canada." *Canadian Journal of Regional Science*. 23(3): 403–426.

Kazemipur, Abdolmohammad. 2004. *An Economic Sociology of Immigrant Life in Canada*. New York: Nova Science Publishers.

Kazemipur, Abdolmohammad, and Shiva Halli. 2000. *The New Poverty in Canada: Ethnic Groups and Ghetto Neighbourhoods*. Toronto: Thompson Educational Publishing, Inc.

Kazemipur, Abdolmohammad, and Shiva Halli. 2001. "Immigrants and 'New Poverty': The Case of Canada." *International Migration Review*. 35(4): 1129–1156.

Knack, Stephen, 2001. "Trust, Associational Life and Economic Performance." Pp. 172–202. *The Contribution of Human and Social Capital to Sustained Economic Growth and Well-Being*. J.F. Helliwell, editor. Human Resources Development Canada.

Knack, Stephen, and Philip Keefer. 1997. "Does Social Capital Have an Economic Payoff? A Cross-Country Investigation." *The Quarterly Journal of Economics*. 112(4): 1251–1288.

La Porta, Rafael, Florencio Lopez-de-Silanes, Andrei Shleifer, and Robert Vishny. 1997. "Trust in Large Organizations." *American Economic Review*. 87(2): 333–38.

Leigh, Andrew. 2006a. "Diversity, Trust and Redistribution." *Dialogue* 25(3): 43–9.

Leigh, Andrew. 2006b. "Trust, Inequality and Ethnic Heterogeneity." *The Economic Record*. 82(258): 268–280.

Letki, Natalia. 2008 (forthcoming). "Does Diversity Erode Social Cohesion? Social Capital and Race in British Neighbourhoods." *The Political Studies*.

Ley, David. 2005. "Post-Multiculturalism?" *Working Paper Series*, No. 05–18. Vancouver Centre of Excellence: Research on Immigration and Integration in the Metropolis.

Lieberson, S. 1981. "An Asymmetrical Approach to Segregation." *Ethnic Segregation in Cities*. Peach, Robinson, and Smith, editors. USA: University of Georgia Press.

Lin, Nan. 1995. "Les Ressources Sociales: Une Theorie du Capital Social." Revue Francaise de Sociologie. 36(4): 685–704.

Lipset, Seymour Martin. 1997. *American Exceptionalism: A Double-Edged Sword*. New York: W. W. Norton.

Lloyd, John. 2006a. "Research Shows Disturbing Picture of Modern Life." *Financial Times* (on-line). October 8.
http://www.ft.com/cms/s/2584c7b6-56ea-11db-9110-0000779e2340.html

Lloyd, John. 2006b. "Harvard Study Paints Bleak Picture of Ethnic Diversity." *Financial Times* (on-line). October 9. http://www.ft.com/cms/s/7e668728-5732-11db-9110-0000779e2340.html

Marschall, Melissa J., and Dietlind Stolle. 2004. "Race and the City: Neighborhood Context and the Development of Generalized Trust." *Political Behavior*. 26(2): 125–152.

Marsden, Peter V. 2005. "The Sociology of James S. Coleman." *Annual Review of Sociology*. 31: 1–24.

Massey, Douglas S., and Nancy A. Denton. 1988. "The Dimensions of Residential Segregation." *Social Forces*. 67(2): 281–315.

Moody, James. 2001. "Race, School Integration, and Friendship Segregation in America." *American Journal of Sociology*. 107(3): 679–716.

Norris, Pippa. 2002. *Democratic Phoenix: Reinventing Political Activism*. Cambridge: Cambridge University Press.

Norris, Pippa, and Ronald Inglehart. 2004. *Sacred and Secular: Religion and Politics Worldwide*. Cambridge: Cambridge University Press.

Norusis, M. J. 1990. *The SPSS Manual*. Chicago, IL: SPSS Inc.

Offe, Clause. 1999. "How Can We Trust Other Fellow Citizens?" *Democracy and Trust*. Mark Warren, editor. Cambridge, UK: Cambridge University Press, Pp. 42–87.

Pettigrew, Thomas F. 1998. "Intergroup Contact Theory." *Annual Review of Psychology*. 49: 65–85.

Pettigrew, Thomas F., and Linda R. Tropp. 2006. "A Meta-Analytic Test of Intergroup Contact Theory." *Journal of Personality and Social Psychology*. 90(5): 751–783.

Policy Research Initiative. 2003. *The Opportunity and Challenge of Diversity: A Role for Social Capital?* Conference in Montreal, Quebec, October 12–25. http://policyresearch.gc.ca/page.asp?pagenm=sc_conf .

Portes, Alejandro. 1995. "Children of Immigrants: Segmented Assimilation and Its Consequences." Pp. 248–279. *The Economic Sociology of Immigration: Essays in Networks, Ethnicity, and Entrepreneurship*. Portes, editor. New York: Russell Sage Foundation.

Portes, A. 1998. "Social Capital: Its Origins and Applications in Modern Sociology." *Annual Review of Sociology*. 24: 1–24.

Powers, Daniel A., and Christopher G. Ellison. 1995. "Interracial Contact and Black Racial Attitudes: The Contact Hypothesis and Selectivity Bias." *Social Forces*. 74: 205–226.

Putnam, Robert D. 1993. *Making Democracy Work: Civic Traditions in Modern Italy*. New Jersey: Princeton University Press.

Putnam, Robert D. 1995. "Tuning In, Tuning Out: The Strange Disappearance of Social Capital in America." *Political Science and Politics*. 27(4): 664–683.

Putnam, Robert D. 2000. *Bowling Alone: The Collapse and Revival of American Community*. New York: Simon and Schuster.

Putnam, Robert D. 2001. "Social Capital: Measurement and Consequences." *ISUMA*. 2(1): 41–51.

Putnam, Robert D., editor. 2002. *Democracies in Flux: The Evolution of Social Capital in Contemporary Society*. New York: Oxford University Press.

Putnam, Robert D. 2003. "Social Capital in a Diverse Society: Who Bridges? Who Bonds?" Presented at the conference, *The Opportunity and Challenges of Diversity: A Role for Social Capital*. Montreal, November 23–25.

Putnam, Robert D. 2007. "*E Pluribus Unum*: Diversity and Community in the Twenty-first Century." The 2006 Johan Skytte Prize Lecture. *Scandinavian Political Studies,* 30 (2): 137–174.

Rice, Tom W., and Jan L. Feldman. 1997. "Civic Culture and Democracy from Europe to America." *Journal of Politics.* 59(4), 1143–1172.

Schoenfeld, Eugen. 1978. "Image of Man: The Effect of Religion on Trust." *Review of Religious Research.* 20(1): 61–67.

Schuman, Howard, and Shirley Hatchett. 1974. *Black Racial Attitudes.* Ann Arbor, Michigan: University of Michigan Institute for Social Research.

Sherif, Muzafar, O.J. Harvey, B.Jack White, William r. Hood, and Carolyn W. Sherif. 1961. *Intergroup Conflict and Cooperation: The Robber's Cave Experiment.* Norman: University of Oklahoma Press.

Sigelman, Lee, Timothy Bledsoe, Susan Welch, and Michael W. Combs. 1996. "Making Contact? Black-White Social Interaction in an Urban Setting." *American Journal of Sociology.* 101(5): 1306–1332.

Sigelman, Lee, and Susan Welch. 1993. "The Contact Hypothesis Revisited: Black-White Interaction and Positive Racial Attitudes." *Social Forces.* 71(3): 781–795.

Skocpol, Theda. 2002. "From Membership to Advocacy." *Democracies in Flux: The Evolution of Social Capital in Contemporary Society.* Putnam, editor. New York: Oxford University Press.

Soroka, Stuart N., John F. Helliwell, and Richard Johnston. 2007. "Measuring and Modelling Trust." *Diversity, Social Capital and the Welfare State.* F. Kay and R. Johnston, editors, *Social Capital, Diversity, and the Welfare State.* University of British Columbia Press.

Statistics Canada. 2004. *Ethnic Diversity Survey Analytical File,* accessed through Prairie Regional Research Data Centre.

Tabachnick, Barbara G., and Linda S. Fidell. 1996. *Using Multivariate Statistics, Third Edition.* New York: HarperCollins College Publishers.

Uslaner, Eric M. 1999. "Democracy and Social Capital." Pp. 121–150. *Democracy and Trust.* M. Warren, editor. Cambridge: Cambridge University Press.

Ulph, Alistair. 2006. "Immigrant Societies Build New Identity." *Financial Times.* October 1. http://www.ft.com/cms/s/e3f48944-58c4-11db-b70f-0000779e2340.html .

Valpy, Michael. 2006. "Why the Global Rage Hasn't Engulfed Canada." *The Globe and Mail.* February 8. Pp. A14

van Laar, Colette, Shana Levin, Stacey Sinclair, and Jim Sidanius. 2005. "The Effect of University Roommate Contact on Ethnic Attitudes and Behaviour." *Journal of Experimental Social Psychology.* 41: 329–345.

Warren, Mark E. 1999. "Introduction." *Democracy and Trust.* M. Warren, editor. Cambridge: Cambridge University Press. Pp. 1–21.

Watts, Duncan J. 2004. "The 'New' Science of Networks." *Annual Review of Sociology.* 30: 243–270.

Weinberg, Steven. 1993. *Dreams of a Final Theory: The Scientific Search for the Ultimate Laws of Nature.* New York: Vintage Books.

White, M.J. 1983. "The Measurement of Spatial Segregation." *American Journal of Sociology.* 88: 1008–1018.

Wilson, William Julius. 1987. *The Truly Disadvantaged: The Inner City, the Underclass, and Public Policy.* Chicago: The University of Chicago Press.

Wilson, William Julius. 2000. *When Work Disappears: the World of the New Urban Poor.* New York: Vintage Books.

Wuthnow, Robert. 1998. *Loose Connections: Joining Together in America's Fragmented Communities.* Cambridge: Harvard University Press.

Wuthnow, Robert. 2002. "Bridging the Privileged and the Marginalized?" *Democracies in Flux.* Putnam, editor. New York: Oxford University Press. Pp. 59–102.

Yancey, George. 1999. "An Examination of the Effects of Residential and Church Integration on Racial Attitudes of Whites." *Sociological Perspectives.* 42(2): 279–304.

Zak, P.J., and S. Knack. 2001. "Trust and Growth." *Economic Journal.* 111: 295–321.

Appendix 2.1: Composite Indexes

The following are a series of SPSS syntax commands used to calculate the composite indexes. Each index is equal to the respondent's scores on the specific variables associated with each component, multiplied by the corresponding factor loading. The values have then been divided by the maximum values that could be acquired for each index, so that all index scores get reported as a fraction of 1, which is the theoretical maximum. This standardizes the scores so they vary between 0 and 1.

- COMPUTE Trust =
  (trt_q110_rec*0.684)+(trt_q310_rec*0.486)+(trt_q330_rec*0.721)+(trt_q390_rec*0.695)+(trt_q400_rec*0.752))/((1*0.684)+(4*0.486)+(4*0.721)+(4*0.695)+(4*0.752))
- COMPUTE Confidence_in_public_institution =
  ((trt_q640_rec*0.631)+(trt_q650_rec*0.635)+(trt_q660_rec*0.640)+(trt_q670_rec*0.621))/((0.631*1)+(0.635*1)+(0.640*1)+(0.621*1))
- COMPUTE Voting =
  ((pe_q110_rec*0.877)+(pe_q120_rec*0.881)+(pe_q130_rec*0.780))/((1*0.877)+(1*0.881)+(1*0.780))
- COMPUTE Religion =
  ((ce_q114_rec*0.781)+(religatt_rec*0.843)+(rl_q105_rec*0.632))/((1*0.781)+(4*0.843)+(4*0.632)).
- COMPUTE Volunteering =
  ((ce_q240_rec*0.334)+(vcg_q300_rec*0.822)+(vcg_q310_rec*0.815))/((0.334*1)+(0.822*1)+(0.815*22.5)).
- COMPUTE Political_Party_Activism=
  ((ce_q111_rec*0.862)+(pe_q230_rec*0.870))/((1*0.862)+(1*0.870)).
- COMPUTE Neighbourliness=
  ((dor_q228_rec*0.879)+(dor_q229_rec*0.868))/((0.879*1)+(0.868*1)).
- COMPUTE Political_information_acquiring_sharing=
  ((pe_q220_rec*0.574)+(pe_q250_rec*0.576)+(pe_q290_rec*0.581)+(q110gd_rec*0.357))/((1*.574)+(1*0.567)+(1*0.581)+(1*0.357)).
- COMPUTE Confidence_private_institution=
  ((trt_q680_rec*0.728)+(trt_q690_rec*0.738))/((0.728*1)+(0.738*1)).
- COMPUTE Group_Activity=
  ((ce_q112_rec*0.742)+(ce_q330_rec*0.502)+(yer_q110_rec*0.570))/((0.742*1)+(0.502*4)+(0.570*1)).
- COMPUTE Political_expression=
  ((pe_q260_rec*0.676)+(pe_q300_rec*0.726))/((0.676*1)+(0.726*1)).

- COMPUTE Social_network=
  ((scf_q100_rec*0.566)+(scf_q120_rec*0.722))/((0.566*20)+(0.722*4)).
- COMPUTE Cultural_Community_participation=
  ((ce_q113_rec*0.519)+(ce_q115_rec*0.671))/((0.519*1)+(0.671*1)).
- COMPUTE Donation_Youth_Business=
  ((trt_q700_rec*0.387)+(vcg_q340_rec*0.462)+(yer_q120_rec*0.656))/((0.387*1)+(0.462*1)+(0.656*1)).
- COMPUTE Self_Interest=
  ((ce_q110_rec*0.455)+(ce_q116_rec*0.502)+(pe_q310_rec*1.609))/((0.455*1)+(0.502*1)+(1.609*3)).

Appendix 7.1: Variables and Data Sources

City	Percentage of City Population who report trusting others	Pop2001_1000s (100% sample)	Pop2001_1000s (20% sample)	Immigrant population as a proportion of city population	Immpop 1000S (20% Sample)	Average annual income ($)	Ethnic Diversity Index (IQV)	Population in private households - Incidence of low income in 2000 %	Economic families - Incidence of low income in 2000 %	Unattached individuals 15+ - Incidence of low income in 2000 %
Prince George	55.7	85.035	84.610	10.0	8.500	30,927	0.87	14.7	11.8	36.9
Nanaimo	66.7	85.664	84.465	16.1	13.615	26,733	0.88	18.8	14.1	43.3
Victoria	72.7	311.902	306.970	19.5	59.965	30,361	0.87	14.4	9.2	36.1
Vancouver	64.7	1,986.965	1967.475	39.0	767.720	31,421	0.90	20.8	17.1	39.8
Abbotsford	58.1	147.370	144.990	22.4	32.420	27,011	0.90	13.5	11.2	36.5
Kamloops	61.6	86.491	86.025	10.9	9.405	28,195	0.89	16.0	11.9	43.4
Kelowna	60.4	147.739	145.950	14.3	20.815	27,773	0.88	13.1	9.3	36.1
Edmonton	61.2	937.845	927.020	18.5	171.050	30,468	0.92	16.2	12.4	39.2
Red Deer	68.1	67.707	66.565	9.0	5.960	30,209	0.86	13.6	9.7	31.4
Calgary	63.4	951.395	943.310	21.7	205.000	35,693	0.92	14.1	10.6	33
Lethbridge	65.1	67.374	66.275	12.3	8.180	26,991	0.91	16.4	11.1	40.7
Medicine Hat	66.6	61.735	61.120	8.0	4.910	27,302	0.81	13.7	10.5	34.3
Saskatoon	63.5	225.927	222.630	8.2	18.240	28,045	0.90	18.0	13.5	40.4
Regina	70.0	192.800	190.020	7.8	14.885	29,780	0.89	15.5	11.5	36.3
Winnipeg	63.1	671.274	661.725	17.0	112.750	28,560	0.94	19.2	14.6	43.7
Thunder Bay	68.9	121.986	120.370	11.2	13.440	29,728	0.91	14.1	10.3	37.2
Sault Ste. Marie	59.3	78.908	77.815	11.5	8.960	26,968	0.85	16.4	13.3	43
Greater Sudbury	60.1	155.601	153.895	7.1	10.955	29,000	0.78	14.9	11.5	41.1
North Bay	70.5	63.681	62.640	5.2	3.260	27,725	0.74	16.7	13.2	42.8
Barrie	59.3	148.480	146.970	11.9	17.550	31,453	0.80	10.0	8.2	31
Sarnia	70.5	88.331	87.460	12.9	11.295	30,754	0.83	13.4	10.2	34.1
Windsor	54.4	307.877	304.955	22.9	69.895	33,938	0.90	13.2	10.4	32.9
London	63.2	432.451	427.215	19.6	83.585	31,050	0.88	15.1	11.4	34.7
Guelph	68.0	117.344	115.775	20.5	23.715	32,909	0.89	9.6	6.7	30.5
Brantford	57.0	86.417	85.125	14.7	12.490	27,675	0.85	15.2	12.2	36
Kitchener	63.9	414.284	409.765	22.6	92.620	32,457	0.89	11.3	8.8	29.2
St. Catherines - Niagara	62.9	377.009	371.405	18.3	67.800	28,693	0.88	13.2	9.9	34.7
Hamilton	58.2	662.401	655.060	24.5	160.240	32,379	0.91	16.7	13.4	41.9
Toronto	55.2	4,682.897	4647.955	45.0	2,091.095	35,618	0.95	16.7	14.4	35.1
Oshawa	55.7	296.298	293.545	15.9	46.760	33,682	0.83	9.4	7.7	28.6
Belleville	49.2	87.395	86.315	9.0	7.785	27,636	0.73	14.8	11.7	35.5
Kingston	61.0	146.838	142.770	12.9	18.425	30,374	0.81	15.2	11.2	38.2
Ottawa-Hull	58.3	1,063.664	1050.755	18.4	193.665	36,608	0.82	15.0	11.6	33.4
Montreal	37.0	3,426.350	3380.645	19.4	656.440	29,199	0.71	22.2	17.4	46.1
Saint-Jean-sue-Richelieu	53.0	79.600	78.600	2.7	2.120	26,101	0.37	16.2	12	42.4
Granby	39.0	60.264	58.770	2.5	1.440	26,398	0.31	14.7	10.6	38.5
Drummondville	45.0	68.451	66.860	1.9	1.270	23,962	0.28	18.2	13.1	46.5
Trois-Rivieres	32.5	137.507	134.645	1.7	2.285	25,185	0.33	19.8	14.4	48.2
Sherbrooke	51.2	153.811	150.385	5.1	7.690	25,493	0.39	18.5	12.6	44.6
Quebec city	37.3	682.757	673.100	3.2	21.710	27,939	0.39	18.8	13.8	46.2
Chicoutimi-Jonquiere	36.3	154.938	153.020	1.0	1.490	25,908	0.32	16.4	13.1	45.2
Saint John	61.4	122.678	121.340	4.1	5.030	26,932	0.69	17.8	15.2	38.8
Moncton	58.5	117.727	115.815	3.1	3.600	26,129	0.70	14.0	10.9	37.8
Halifax	61.6	359.183	355.945	7.4	26.340	29,586	0.74	15.5	11.9	36.5
St.John's	59.8	172.918	171.105	3.1	5.345	27,061	0.65	17.4	14.7	44.2

- *Percentage of City Population who report trusting others:* Proportion of the population who responded that 'people could be trusted'.
- *Pop2001_1000s (100% sample):* Information based on whole population.
- *Pop2001_1000s (20% sample):* Estimation based on the 20% sample for whole population.
- *Immpop_(1000s):* Estimate of the total number of immigrants in each city; 'Non-immigrant population (from 20% sample data) was subtracted from the variable 'Total population by immigrant status and place of birth (20% sample').
- *Immigrant population as a proportion of city population:* Percent of city population composed of immigrants. Variable was computed by dividing the estimated number of immigrants in each city (Immpop_1000s) by the estimated (20% sample) population for each city.
- *Average annual income:* Variable for average income per CMA.
- *Ethnic Diversity Index (IQV):* The variable 'Total population by ethnic origin' contained 61 possible responses for ethnic origin. All of these were included in the calculation. 'Single responses' were utilized.
- *Population in private households – Incidence of low income in 2000 (%):* Percent of population falling below low income cut off for private households.
- *Economic families – Incidence of low income in 2000 (%):* Percent of population falling below low income cut off for economic families.
- *Unattached individuals 15+ – Incidence of low income in 2000 (%):* Percent of population falling below low income for unattached individuals.

The variable on Trust is based on the Canadian General Social Survey (GSS), Cycle 17, Statistics Canada, 2003. The masterfile was made available to the researcher through the Prairie Regional Research Data Centre, University of Calgary. The rest of variables were accessed through the 2001 Canadian Census, Canadian Census Profile, Census Tract level accessed through the Computing in the Humanities and Arts and Sciences Centre (CHASS), University of Toronto. Available at: http://www.chass.utoronto.ca/.

Appendix 7.2

	Correlations between generalized trust and selected variables for major Canadian cities.		
	Pearson Correlation	Sig. (2-tailed)	N
Pop, 2001 - 100%	-0.19	(0.21)	45
Average income	**0.32**	(0.03)	45
Incidence Low Income for Economic Families	**-0.36**	(0.02)	45
Incidence Low Income for Population in Private Households	**-0.37**	(0.01)	45
Incidence Low Income for Unattached Individuals	**-0.49**	(0.00)	45
Ethnic Diversity (IQV)	**0.75**	(0.00)	45
Immigrant population (Percent based on 20% sample)	0.29	(0.05)	45
Source: *Trust - GSS 17 Masterfile* Others - 2001 Census Profile CMA-CA-CT Level			

Appendix 7.3: IQV

The degree of ethnic diversity has been calculated using the Index of Qualitative Variation (IQV). This index measures the existing diversity of a community (for example, a city), and reports it as a number that varies between 0 and 1, with 0 showing the state of no diversity, and 1 the state of maximum diversity (Frankfort-Nachmias and Leon-Guerrero, 2006). The formula for IQV, shown below, takes into account the overall number of people in each city (N), the number of possible categories (K) for the feature under study (here, single ethnic origins), as well as the number of people in each category (Fi and Fj).

$$IQV = \frac{\sum f_i f_j}{\frac{K(K-1)}{2} \left(\frac{N}{K}\right)^2}$$

Appendix 11.1: Dissimilarity Index (DI)

The most widely used measure of residential evenness is the Dissimilarity Index, which has the following formula:

$$DI = \frac{1}{2} \sum_{i=1}^{n} \left| \frac{N_{1i}}{N_1} - \frac{N_{2i}}{N_2} \right|$$

Where N_{1i} = population of group 1 in *ith* census tract, N_{2i} = population of group 2 in *ith* census tract, N_1 = total population of group 1 in city, and N_2 = total population of group 2 in city. The value of D is equal to the proportion of the minority (or majority) population which would have to be redistributed so that each parcel would have exactly the same composition as the city as a whole (White, 1983: 1009).

Appendix Figure 11.1: The Graphical Representation of DI

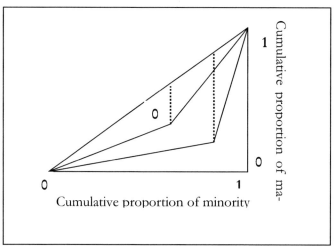

To illustrate the meaning of DI graphically, James and Taeuber (1985: 6-7) have suggested Appendix Figure 11.1, in which the cumulative proportion of minority group X is plotted against the cumulative proportion of the majority proportion. The condition of zero segregation is indicated by the diagonal line. The line for a completely segregated population would lie along the x-axis from 0 to 1 and then rise along the y-axis (James and

Taeuber, 1985: 6). For the cases between the two extremes, "the D Index is the maximum distance between the segregation curve and the diagonal" p. 6 (i.e., D1 and D2).

Using a hypothetical data set (see the table below), Lieberson (1981) shows a step-by-step way of calculating DI. In this table, columns 2 and 3 express the number of the Blacks and Whites in each sub-area, and columns 5 and 6, the percentage of black and white populations who live in each sub-area. By converting to percentage distributions, the DI intentionally ignores the absolute numbers involved in each group, and simply compares the two percentage distributions to determine how similar they are to each other. This is done by summing differences between the Black and White percentages in each sub-area (ignoring signs). The DI is one-half the sum of these differences. In the example at hand, the index is

$$(|30-37| + |10-29| + |0-30| + |60-4|) / 2 = 56 \text{ (on a scale of 0 to 100)}.$$

The DI ranges from 0 (in which case there is no segregation because the percentage distribution is identical for each group) to 100 (which would occur in the case of maximum segregation, such that Blacks would be found only in sub-areas where Whites were absent and vice versa) (Lieberson, 1981). DI has been also called the "displacement index," as it may be interpreted as the proportion of the minority population who would have to change their tract of residence to make the minority and/or majority ratio in each tract equal to the overall ratio (Duncan and Duncan, 1955: 211).

A Hypothetical Data Set For Computation of Segregation Index

Subareas	Number: Blacks	Number: Whites	Total	Percent: Black	Percent: White	Black Proportion of Subarea Total
A	60	370	430	30	37	0.14
B	30	290	310	10	29	0.065
C	0	300	300	0	30	0
D	120	40	160	60	4	0.75
SUM	200	1000	1200	100	100	

Source: Lieberson 1981: 62

DI has its shortcomings. Regarding the interpretation of DI as the proportion of the minority population that needs to be transferred in order to reach evenness, for example, Massey and Denton (1988) point out that these transfers of minority members need to be only from overrepresented to underrepresented areas, in order to make a difference in the value of the DI.

The other weakness of the DI is what was once considered to be its advantage, that is, its immunity from "compositional" influences. Referring to the preceding table, one can easily see that the Dissimilarity Index will remain intact if the white population, for example, were one-tenth of its original size, but had the same distribution: in other words, if the number of whites in each sub-area was divided by 10, that would not cause any change in the DI. This feature is not always desirable, especially if one needs to measure the level of interaction between different groups (Lieberson, 1981). It follows that, when the number of minority members is small, the DI is not very reliable (Massey and Denton, 1988).

Appendix 12.1: Logistic Regression

Logistic Regression (LR), useful for identifying group membership, estimates the probability of an event occurring (Norusis, 1990: B29). LR is a very robust and flexible technique due to the fact that, unlike many other multivariate techniques, it does not require any assumptions about the distributions of the predictor variables. This means that in logistic regression, predictors do not have to be normally distributed, linearly related, or of equal variance within each group (Tabachnick and Fidell 1996: 575). It is most appropriate to use when the dependent variable is dichotomous, e.g., group membership or non-membership. What logistic regression can tell us is the probability of a particular outcome for each case, and the extent to which a particular predictor increases or decreases the odds of that outcome. In the present study, the dependent variable, trust, is of such a dichotomous nature – consisting of 'yes' and 'no' answers, trusting and non-trusting.

Halli and Rao (1992) have extensively discussed the mathematical properties of logistic regression. Briefly, the logistic regression technique operates on the individual or micro-level rather than on aggregated data, and is analogous to linear regression in that a continuous response variable is modeled as a linear function of a set of continuous predictors. The technique assumes that each member of the population has some underlying probability of falling in the category of interest in the dependent variable, as a function of a set of given independent variables. Therefore, in the population, each member with a given set of characteristics has a P chance of being in the category of interest in the dependent variable (here, trusting) and 1-P chance of being in the opposite category (non-trusting).

Let P_i be the probability that the *ith* person in the sample is in the category of interest (trusting), and $(1-P_i)$ be the probability that he or she is in the opposite category (non-trusting). Clearly, $P_i/(1-P_i)$ equals the odds of being in the category of interest for the *ith* individual. Now $\log(P_i/(1-P_i))$, the log of the odds of being in the category of interest, is a continuous variable that theoretically can take on any value in the range $(-\infty, +\infty)$. Also, let $X_{i1}, X_{i2}, \ldots, X_{iK}$ be a set of K continuous predictor variables measured on the ith individual in the sample. Then the logistic regression model for the log odds, given a particular vector of scores on the K predictor variables, is

$$\ln\frac{P_i}{1-P_i} = \beta_0 + \beta_1 X_{i1} + \beta_2 X_{i2} + \ldots + \beta_k X_{ik}$$

And the corresponding multiplicative model for the odds is

$$\ln\frac{P_i}{1-P_i} = \beta_0 + \beta_1 X_{i1} + \beta_2 X_{i2} + \ldots + \beta_k X_{ik}$$

Estimates of the betas, or regression coefficients, in these two equations are obtained by the method of maximum likelihood. These coefficients are then used to estimate the relative contribution of each independent variable. There are a couple of alternative ways to interpret the output produced by logistic regression. The one adopted here relies on the magnitude of Exp(B). The magnitude of Exp(B) in our model determines the contribution of each variable in the increase or decrease of the odds of being trusting. The values smaller than 1 indicate a negative impact, while the values greater than 1 a positive one.